Invertebrate Cell System Applications

Volume I

Editor

Jun Mitsuhashi

Professor, Faculty of Agriculture
Tokyo University of Agriculture and Technology
Tokyo, Japan

CRC Press
Taylor & Francis Group
Boca Raton London New York

CRC Press is an imprint of the
Taylor & Francis Group, an **informa** business

First published 1989 by CRC Press
Taylor & Francis Group
6000 Broken Sound Parkway NW, Suite 300
Boca Raton, FL 33487-2742

Reissued 2018 by CRC Press

Library of Congress Cataloging-in-Publication Date

Invertebrate cell system application.
    Bibliography: p.
    Includes index.
    1. invertebrates--Cultures and culture media.
2. Cell culture.   I. Mutsuhashi, Jun.
QL362.81579 1989 592' .0072'4 88-19379
ISBN 0-8493-4373-9 (v. 1)
ISBN 0-8493-4374-7 (v. 2)

A Library of Congress record exists under LC control number: 88019379

ISBN 13: 978-1-315-89474-4 (hbk)
ISBN 13: 978-1-351-07384-4 (ebk)

Visit the Taylor & Francis Web site at http://www.taylorandfrancis.com and the
CRC Press Web site at http://www.crcpress.com

PREFACE

A quarter century has elapsed since Dr. T. D. C. Grace established the first continuous cell lines from insects. When he reported the establishment in 1962, I had just started insect cell culture studies at Dr. K. Maramorosch's laboratory at Boyce Thompson Institute, Yonkers, NY. His result encouraged those who had been attempting to get continuously growing cells from insects. We had performed extensive studies on insect cell cultures with special reference to plant virus replication in cultured leafhopper vector cells, and obtained some successful results. Among them, I, in collaboration with Dr. Maramorosch, have published the formulation of a medium called MM (Mitsuhashi and Maramorosch) medium now, which later expanded its use to mosquito cell cultures by the late Dr. K. R. P. Singh and is now being used for the studies of arboviruses in cultured mosquito cells in many laboratories of the world.

Reports on establishment of invertebrate cell lines have increased since 1970. Now, more than 200 cell lines of invertebrates exist, including unpublished ones. At the early stages of invertebrate cell cultures, most efforts had been concentrated to growing cells. During those stages there were many difficulties for culturing invertebrate cells, and overcoming these difficulties was immediately accepted as a new contribution to science. However, recently, obtainment of new cell lines became common, and the papers reporting only new cell lines are hardly accepted as scientific papers. The cell culture itself is a technique and the resultant cell lines are materials for experiments in various scientific fields. Now we are in the time of application of invertebrate cell lines to various sciences and technology. However, existing cell lines are not enough to satisfy these demands. Especially for invertebrates other than Arthropoda, only one continuous cell line from a snail is available. There are still many groups of organisms in which cell cultures are not successful. Therefore, we have to continue to work further on the culture techniques.

The aim of these two volumes is to supply information about invertebrate cell cultures and their applications to various fields of sciences. The sections of physiology, biochemistry, and endocrinology of cultured cells will give valuable information not only for research work in these fields, but also for improvement of culture techniques. The sections of biotechnology and molecular biology of invertebrate cells are relatively new fields of sciences, and will give up-dated techniques and information to readers. The section of pathology is divided into three parts according to groups of host organisms. Some articles of those sections are deeply related to biotechnology and molecular biology. The last section provides information about new cell lines. Some of them were derived from animals in which cell cultures had been considered extremely difficult if not impossible. The list of invertebrate cell lines is very useful when one plans to use invertebrate cells in his experiments. The list will assist the applications of invertebrate cell lines to various studies. The books will be of interest to those working in the fields of physiology, genetics, endocrinology, biochemistry, molecular biology, biotechnology, virology, parasitology, microbiology, entomology, and so on.

I deeply express my thanks to Dr. K. Maramorosch of Rutgers University for his recommendation, encouragement, and invaluable suggestions to edit these two volumes. I also appreciated very much the cooperation of all the contributors by preparing their excellent manuscripts in a timely manner.

I hope that these two volumes will satisfy the requirements of those who are interested in invertebrate cell cultures.

Jun Mitsuhashi

THE EDITOR

Jun Mitsuhashi is Professor of Applied Entomology at the Tokyo University of Agriculture and Technology, Fuchu, Tokyo 183, Japan.

Dr. Mitsuhashi graduated in 1955 from Faculty of Agriculture, University of Tokyo, Tokyo, with a degree of Bachelor of Agriculture. Thereafter he worked on insect endocrinology and cell cultures at the National Institute of Agricultural Sciences. He worked with Dr. K. Maramorosch at Boyce Thompson Institute, Yonkers, NY from 1962 to 1964 by the fellowship of the Agency of Science and Technology, Japan, and of the National Science Foundation, U.S. In 1965 he obtained his degree of Doctor of Agriculture from the University of Tokyo. He also worked with Dr. T. D. C. Grace on insect cell cultures at the Division of Entomology, Commonwealth Scientific and Industrial Research Organization, Canberra from 1968 to 1969. In 1984 Dr. Mitsuhashi was appointed chief of the Laboratory of Insect Pathology at the Forestry and Forest Product Research Institute, Tsukuba, Japan, and in 1988 he was appointed Professor of Applied Entomology at the Tokyo University of Agriculture and Technology. He is currently teaching applied entomology and insect physiology to students and is organizing postgraduate courses.

Dr. Mitsuhashi is a member of the Tissue Culture Association, Society for Invertebrate Pathology, Zoological Society of Japan, Japanese Tissue Culture Association, Japanese Society of Applied Entomology and Zoology, and Japanese Forestry Society.

He has received awards from the Japanese Society of Applied Entomology and Zoology in 1968 and from Japanese Society for Agronomy in 1980 for his works on insect endocrinology and insect tissue cultures.

He has presented over ten invited lectures at international meetings, and also over ten guest lectures at universities and institutes. He has published more than 80 research papers and more than 20 review papers. His current major research interests include the physiology and biochemistry of cultured insect cells.

CONTRIBUTORS

Phillip M. Achey
Professor
Department of Microbiology and Cell
  Science
University of Florida
Gainesville, Florida

Noriaki Agui
Chief
Department of Medical Entomology
National Institute of Health
Tokyo, Japan

Jacqueline Becker
Technicienne
U. A. CNRS 1135
Universite P. et M. Curie
Paris, France

Edward Berger
Professor
Department of Biology
Dartmouth College
Hanover, New Hampshire

Martin Best-Belpomme
Research Director
U. A. CNRS 1135
Universite P. et M. Curie
Paris, France

Philippe Beydon
Assistant - Agrege
Department of Biology
Ecole Normale Supérieure
Paris, France

Catherine Blais
Department of Biology
Ecole Normale Supérieure
Paris, France

Ethel Brandt
Entomologist
Insect Reproduction Laboratory, ARS
United States Department of Agriculture
Beltsville, Maryland

Anne-Marie Courgeon
Research
U. A. CNRS 1135
Universite P. et M. Curie
Paris, France

Walter Doerfler
Professor
Institute of Genetics
University of Cologne
Cologne, West Germany

Guy Echalier
Professor
U. A. CNRS 1135
Universite P. et M. Curie
Paris, France

A. M. Fallon
Associate Professor
Department of Entomology
University of Minnesota
St. Paul, Minnesota

Haruhiko Fujiwara
Department of Technology
Laboratory of Radiation Biology
National Institute of Health
Tokyo, Japan

Ronald H. Goodwin
Research Entomologist
Rangeland Insect Laboratory
United States Department of Agriculture
Montana State University
Bozeman, Montana

Linda A. Guarino
Research Scientist
Department of Entomology
Texas A & M University
College Station, Texas

B. Happ
Institute of Genetics
University of Cologne
Cologne, West Germany

George T. Harvey
Research Scientist
Great Lakes Forestry Centre
Canadian Forestry Service
Sault Ste. Marie, Ontario, Canada

Charlotte Hauser
Visiting Scientist
Department of Biology
Massachusetts Institute of Technology
Cambridge, Massachusetts

Kiyoshi Hiruma
Research Scientist
Department of Zoology
University of Washington
Seattle, Washington

Carlo M. Ignoffo
Laboratory Director
United States Department of Agriculture
Agricultural Research Service
Biological Control of Insects Research
 Lab
Columbia, Missouri

Eric Brian Jang
Research Entomologist
Agricultural Research Service
United States Department of Agriculture
Hilo, Hawaii

H. Johansen
Department of Molecular Genetics
Smith Kline & French Laboratories
King of Prussia, Pennsylvania

Johannes Kaiser
Department of Entomology
ETH-Zurich
Zurich, Switzerland

Akihiko Kiyota
Department of Technology
Laboratory of Culture Media Studies
National Institute of Health
Tokyo, Japan

Hans-Dieter Klenk
Professor
Institute of Virology
Philipps-Universitat
Marburg, West Germany

Kazumichi Kuroda
Institute of Virology
Philipps-Universitat
Marburg, West Germany

Yukiaki Kuroda
Professor and Chief
Department of Ontogenetics
National Institute of Genetics
Mishima, Shizouka, Japan

Timothy J. Kurtti
Assistant Professor
Department of Entomology
University of Minnesota
St. Paul, Minnesota

René Lafont
Professor
Department of Biology
Ecole Normale Supérieure
Paris, France

Marcia J. Loeb
Research Physiologist
Insect Reproduction Laboratory, ARS
United States Department of Agriculture
Beltsville, Maryland

Susumu Maeda
Assistant Professor
Department of Entomology
University of California Davis
Davis, California

Hideaki Maekawa
Laboratory Chief
Department of Technology
Laboratory of Radiation Biology
National Institute of Health
Tokyo, Japan

Claude Maisonhaute
Research
U. A. CNRS 1135
Universite P. et M. Curie
Paris, France

Edwin P. Marks
Research Cooperator
Department of Biochemistry and
  Molecular Genetics
Agricultural Research Service
United States Department of Agriculture
Fargo, North Dakota

Horace M. Mazzone
Microbiologist
Northeastern Forest Experimental Station
United States Department of Agriculture
  Forest Service
Hamden, Connecticut

Arthur H. McIntosh
Research Microbiologist
United States Department of Agriculture
Agricultural Research Service
Biological Control Insects Research Lab
Columbia, Missouri

Jun Mitsuhashi
Professor
Faculty of Agriculture
Tokyo University of Agriculture and
  Technology
Fuchu, Tokyo, Japan

Naoko Miyajima
Department of Technology
Laboratory of Radiation Biology
National Institutes of Health
Fuchu, Tokyo, Japan

Jean-Francois Modde
Ecole Normale Supérieure
Paris, France

T. Müller
Institute of Genetics
University of Cologne
Cologne, West Germany

Ulrike G. Munderloh
Research Associate
Department of Entomology
University of Minnesota
St. Paul, Minnesota

Osamu Ninaki
Researcher
Department of Insect Genetics and
  Breeding
National Institiute of Sericultural and
  Entomological Science
Yamanashi, Japan

C. Oellig
Institute of Genetics
University of Cologne
Cologne, West Germany

Teru Ogura
Department of Molecular Genetics
Institute for Medical Genetics
Kumamoto University Medical School
Kumamoto, Japan

Lynn M. Riddiford
Professor
Department of Zoology
University of Washington
Seattle, Washington

Emmanuelle Rollet
Student
U. A. CNRS 1135
Universite P. et M. Curie
Paris, France

Michele Ropp
Student
U. A. CNRS 1135
Universite P. et M. Curie
Paris, France

Martin Rosenberg
Vice President of Biopharmaceutical
  Research & Development
Smith Kline & French Laboratories
King of Prussia, Pennsylvania

Rudolf Rott
Professor
Institut fur Virologie
Justus-Leibig-Universitat
Giessen, West Germany

Karen M. Rudolph
Graduate Student
Department of Biology
Dartmouth College
Hanover, New Hampshire

Yutaka Shimada
Professor
School of Medicine
Chiba University
Chiba, Japan

S. S. Sohi
Research Scientist
Forest Pest Management Institute
Canadian Forestry Service
Sault St. Marie, Ontario, Canada

David A. Stock
Associate Professor
Department of Biology
Stetson University
Deland, Florida

Randel Stolee
Surgical Resident
Department of Surgery
Gundersen Medical Foundation
La Crosse, Wisconsin

Raymond W. Sweet
Senior Investigator
Department of Molecular Genetics
Smith Kline & French Laboratories
King of Prussia, Pennsylvania

Naoko Takada
Department of Technology
Laboratory of Radiation Biology
National Institute of Health
Tokyo, Japan

Ariane van der Straten
Postdoctoral Fellow
Department of Molecular Genetics
Smith Kline & French Laboratories
King of Prussia, Pennsylvania

Dirk F. Went
Institute of Animal Science
ETH-Zurich
Zurich, Switzerland

Charles W. Woods
Research Chemist
Insect Reproduction Laboratory, ARS
United States Department of Agriculture
Beltsville, Maryland

TABLE OF CONTENTS

Volume I

PHYSIOLOGY AND BIOCHEMISTRY OF CULTURED CELLS

ENDOCRINOLOGY IN INVERTEBRATE TISSUE CULTURE

TABLE OF CONTENTS

Volume II

APPLICATIONS OF CELL CULTURES TO INSECT PATHOLOGY

DEVELOPMENT OF NEW CELL LINES

Insects

Invertebrates Other Than Insects

Established Cell Lines

Physiology and Biochemistry of Cultured Cells

Chapter 1

NUTRITIONAL REQUIREMENTS OF INSECT CELLS *IN VITRO*

Jun Mitsuhashi

TABLE OF CONTENTS

I. INTRODUCTION

Nutritional requirements of insect cells cultured *in vitro* have attracted the attention of many investigators in this field for a long time. During the last decade many continuous cell lines were obtained. Among them were some cell lines which had been considered very difficult to culture, if not impossible. Success of obtaining such cell lines seems most likely to be due to the improvement of culture media. In other words, the modification of the culture media satisfied nutritional requirements of the cells for multiplication. There are still many insect species from which cell lines cannot be obtained. In order to develop culture media which can support the cell growth of such insects, information on the nutritional requirements of the cells will give a clue.

Unfortunately, nutritional requirements of the cultured insect cells have not been studied extensively, and the available information is limited. In this article, the author has tried to summarize the information on nutritional requirements of insect cells hitherto obtained.

II. INORGANIC SALTS

Inorganic salts are important to maintain certain ion balance and osmotic pressure. However, very little has been determined about the essentiality of individual salts. According to Gottschewski,[1] Ca^{2+}, Mg^{2+}, and Na^+ were essential for the culture of a fruit fly, *Drosophila melanogaster*, imaginal disks, while Cl^- and $H_2PO_4^+$ were necessary only for the cultures of certain tissues. It has been known that insect cells *in vitro* are flexible to the ionic conditions. Most insect cell culture media have ionic composition of either type A or type B, shown in Table 1. A representative of type A medium is MM medium,[2] and that of type B is Grace's medium.[3] Many insect cell lines can multiply in either of them.[4] This shows that growth of many insect cell lines is not severely affected by the Na/K ratio. Examples of such cell lines are shown in Table 2 with the range of Na/K which permits their growth. From these facts, it can be said that the Na/K ratio is not important for insect cells in cultures.

However, it is also true that some insect cells require a strict Na/K ratio for their growth *in vitro*. Ting and Brooks[5] obtained active cell growth from German cockroach (*Blattera germanica*) embryos only when the Na/K ratio of the medium was 10.0, which was approximately equal to that of hemolymph of the same species. Krause et al.[6] also stressed that a 0.2:1 Na/K ratio in molar concentration was essential for the growth of commercial silkworm (*Bombyx mori*) ovarian cells, although Wyatt[7] has reported that the growth of *B. mori* ovarian cells was not affected by the change of Na/K ratio in Trager's medium.[8] Dependency on the Na/K ratio may be changed by the composition of other components of the medium.

Variations in inorganic salt concentrations result in the variation of osmotic pressure of the media. Most insect cell lines have tolerance to a wide range of osmotic pressure (Table 3).

Some investigators incorporated trace metals in their media. These metals are usually Fe, Cu, Mn, Co, Zn, Mo, etc. Adams et al.[9] have detected Si, P, S, and Cl in the nucleus of the cultured insect cells by X-ray microanalysis. However, the significance of using these trace elements is not certain. $AlCl_3$ and $ZnSO_4 \cdot 7H_2O$ reportedly enhanced the cellular adhesiveness and growth in cultures of the fall armyworm (*Spodoptera frugiperda*) cell line, IPL-21.[10]

III. SUGARS

Sugars are important as an energy source. However, most insect cell lines seem to be able to multiply by using glucose as the sole source of sugars. This is supported by the fact

Table 1
TWO TYPICAL COMPOSITIONS OF INORGANIC SALTS IN INSECT CELL CULTURE MEDIA (mg/100 ml)

Salts	Type A	Type B
NaCl	700	—
$NaH_2PO_4 \cdot 2H_2O$	20	114
$NaHCO_3$	12	35
KCl	20	224
$MgCl_2 \cdot 6H_2O$	10	228
$MgSO_4 \cdot 7H_2O$	—	278
$CaCl_2$	20	100
Na/K ratio	26.8	0.2

Table 2
RANGE OF Na/K VALUES IN WHICH INSECT CELLS CAN GROW

Cell lines	Origin	Na/K ratio	Ref.
A.e.	*Antheraea eucalypti* ovaries	0.67:1.38	25
IPLB-SF-21	*Spodoptera frugiperda* ovaries	0.67:1.38	25
MC-Hz-1	*Heliothis zea* ovaries	0.4:10.5	92
AC-20	*Agallia constricta* embryos	0.2:20	93, 94
Kc	*Drosophila melanogaster* embryos	1:10	95
BT-EP	*Ephedrus plagiator* embryos	0.25:0.5	96, 97
NIAS-AeAl-2	*Aedes albopictus* neonate larvae	0.03:45.6	4
NIH-SaPe-4	*Sarcophaga peregrina* embryos	0.03:45.6	4
Several lepidopteran cell lines (from ovaries, hemocytes, and fat bodies)		0.03:45.6	4

Table 3
FLEXIBILITY OF INSECT CELLS TO OSMOTIC PRESSURES

Cell lines	Range of osmotic pressure which permits cell growth (mOsm/kg)	Ref.
Spodoptera frugiperda (IPL-2-AEIII)	360—375	98
Heliothis zea (IMC-Hz-1)	230—380	92
Drosophila melanogaster (Kc)	225—400	95
Choristoneura fumiferana (IPRI-CF-12)	300—400	99
Manduca sexta (FPMI-MS-12)	250—400	99
Orgyia leucostigma (IPRI-OL-13)	200—488	99

that many insect cell lines can be continuously cultured in serum-free MM medium which contains glucose as the sole carbohydrate. However, some sugars other than glucose have also been known to be utilized by insect cells.

Glucose, fructose, mannose, maltose, trehalose, and sucrose are reportedly utilized by insect cells (Table 4). Although Grace and Brzostowski[11] reported that emperor gum moth (*Antheraea eucalypti*) cells consumed 23% sucrose in the medium, generally sucrose has been said not to be utilized by insect cells nor by vertebrate cells. Recently, however, several cell lines were cultured successfully in a medium which contained sucrose as the sole added carbohydrate.[12] Furthermore, it has been shown that some insect cells liberate α-glucosidase,

Table 4

UTILIZATION OF CARBOHYDRATES BY INSECT CELLS OR TISSUES

Carbohydrates	*T. ni* cells[a]	*A. eucalypti* cells[b]	*D. melanogaster* imaginal disk[c]	*C. erythrocephala* Malpighian tube[d]
Glucose	+	+	+	+
Fructose	+	+	+	+
Mannose	+		+	+
Maltose	+		?	+
Trehalose	+	.		+
Arabinose			−	
Galactose	−		?	+
Ribose			−	
Sorbose	−		−	
Sucrose	−	+	−	
Lactose	−		−	−
Melibiose	−			
Cellobiose	−			
Turanose	−			
Xylose			−	−
Starch			−	
Glycogen			−	

Note: + = consumed; − = not consumed; ? = not determined.

[a]   *Trichoplusia ni* cell line.[100]
[b]   *Antheraea eucalypti* cell line.[11]
[c]   *Drosophila melanogaster* cell line.[1]
[d]   *Calliphora erythrocephala* Malpighian tube.[101]

Table 5

DECREASE OF SUCROSE IN CULTURE MEDIA

Media	Duration of culture (d)	Sucrose conc. (mg/100 ml)	Decrease of sucrose (mg/100 ml)
Unused MTCM-1601	0	673.3 + 6.1	—
NIAS-MB-32[a] cultured MTCM-1601	16	163.5 + 0.7	509.8
NIAS-MaBr-93[b] cultured MTCM-1601	10	644.5 + 0.7	28.8
SES-MaBr-4[c] cultured MTCM-1601	13	595.5 + 3.5	77.8

[a]   The cell line derived from the cabbage armyworm (*Mamestra brassicae*) ovaries.
[b]   The cell line derived from *M. brassicae* hemocytes.
[c]   The cell line derived from *M. brassicae* fat bodies.

which converts 1 mol of sucrose to 1 mol of glucose and 2 mol of fructose, in culture media.[12,13] Therefore, as the result of culturing such cell lines, concentration of sucrose in the media decreased (Table 5).

Maltose also has been used once as the sole added sugar for culturing silkworm ovarian cells.[8]

IV. AMINO ACIDS

Amino acids are necessary to constitute protein. However, not all of the 20 amino acids

Table 6
REQUIREMENTS OF AMINO ACIDS BY INSECT CELL LINES

Amino acids	Insect cell lines					
	EPa[a]	PX-58[b]	MB-19[c]	MaBr-85[d]	CmC$_1$[e]	AA-P[f]
α-Alanine	−	−	−	−	−	−
β-Alanine	−	−	−	−	−	−
Arginine	+	+	+	−	+	+
Aspartic acid	*	−	−	−	−	−
Asparagine	−	−	−	−	−	−
Cystine	+	+	+	+	+	+
Glutamic acid	*	−	−	−	−	−
Glutamine	−	+	+	−	+	+
Glycine	+	−	−	−	−	−
Histidine	+	+	+	+	+	+
Isoleucine	+	+	+	+	+	+
Leucine	+	+	+	−	+	+
Lysine	+	+	+	−	+	+
Methionine	+	+	+	+	+	+
Phenylalanine	+	+	+	−	+	+
Proline	+	+	+	−	+	+
Serine	+	+	+	−	+	+
Threonine	+	+	+	+	+	+
Tryptophan	+	+	+	−	+	+
Tyrosine	+	+	+	−	+	+
Valine	+	+	+	−	+	+

Note: − = nonessential amino acids; + = essential amino acids; * = either of them required.

[a]   EPa: the cell line derived from *Periplaneta americana* embryos.[34]
[b]   PX-58: the cell line derived from *Papilio xuthus* ovarioles.[102]
[c]   MB-19: the cell line derived from *Mamestra brassicae* ovarioles.[103]
[d]   MaBr-85: the cell line derived from *M. brassicae* fat bodies.[104]
[e]   CmC$_1$: the cell line derived from *Culex molestus* ovarioles.[105]
[f]   AA-P: the cell line derived from *Aedes aegypti* neonate larvae.[105]

From Mitsuhashi, J., *Adv. Cell Culture*, 2, 133, 1982. With permission.

which are known to be constituents of protein are essential. Some amino acids can be synthesized by insect cells.

Essentiality of amino acids should be determined by the culturing cells in protein- or peptide-free medium. However, such medium has not yet been developed. By the use of media containing proteins or peptides, only a part of essential amino acids can be determined; and nonessential amino acids cannot be determined because, if one omits a certain amino acid from the medium components, the amino acid may be supplied by the degradation of proteins or peptides.

The essential amino acids are somewhat different in different cell lines. About 14 amino acids are usually essential for the growth of insect cells. They are listed in Table 6. There are ten known essential amino acids for vertebrate or insect cells. On the other hand, putative nonessential amino acids for most insect cells are α-alanine, β-alanine, asparagine, aspartic acid, glutamic acid, and glycine. These six amino acids can be eliminated from culture media altogether without deleterious effects on the growth of the cell lines from the common swallow tail (*Papilio xuthus*), the cabbage armyworm (*Mamestra brassicae*), the Japanese cellar mosquito (*Culex molestus*), and *Aedes aegypti*.[14] However, when these amino acids are present in culture media, they are usually consumed to a greater extent.[15] β-Alanine is

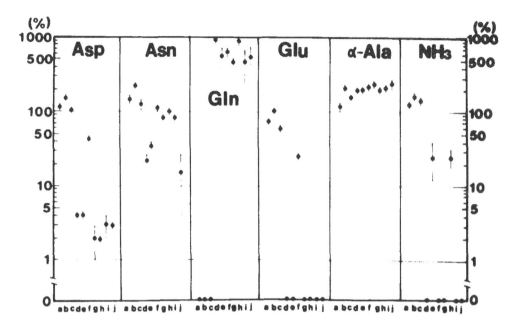

FIGURE 1.   Percent change of acidic amino acids, α alanine, and ammonia in media after 7-d cultures. a, NIAS-MB-19; b, NIAS-MB-25; c, NIAS-MB-32; d, NIAS-MaBr-92; e, NIAS-MaBr-93; f, NIAS-MaBr-85; g, SES-MaBr-1; h, SES-MaBr-2; i, SES-MaBr-3; and j, SES-MaBr-4. The cell lines a through c were derived from pupal ovarioles of *Mamestra brassicae*; d and e, the cell lines derived from larval hemocytes of the same species; f through j, the cell lines derived from larval fat bodies of the same species. Solid circles show averages, and vertical lines show the ranges of standard deviations. (Modified from Mitsuhashi, J., *Appl. Entomol. Zool.*, 22, 533, 1987. With permission.)

somewhat detrimental to the growth of insect cells in most cases.[14,16] Glutamine is essential for many insect cell lines, but not for some others.[15] The cell lines which require glutamine consume it to a greater extent, whereas the ones which do not require glutamine produce it (Figure 1). The latter cell lines consume glutamic acid and free ammonia in the media markedly (Figure 1), and, therefore, glutamine seems to be synthesized from these two substances by the help of glutamine synthetase.[17] The cell lines which require glutamine may lack this enzyme system, and glutamic acid and free ammonia are accumulated in the media during cultivation. It is interesting that the ability of synthesizing glutamine is specific to the tissue from which the cell lines originated. In *M. brassicae*, the cell lines derived from hemocytes and fat bodies have this ability, but the one from ovaries does not (Figure 1).[17]

Studies on consumption of amino acids in media by cultured cells have been made by several investigators. From these results, essentiality of amino acids cannot be determined, but they provide useful information for improvement of the culture media. The patterns of amino acid consumption vary between different cell lines. In general, asparagine, aspartic acid, cystine, glutamine, glutamic acid, and methionine are consumed by cells to a greater extent than other amino acids. α-Alanine is accumulated in media by the culture of any cell line.[15]

V. VITAMINS

For the determination of essentiality of vitamins, the use of chemically defined media is prerequisite. This is because sera and other natural substances which are incorporated into media sometimes contain considerable amounts of vitamins. However, chemically defined

Table 7
VITAMIN REQUIREMENTS IN THE CULTURES OF THE IMAGINAL DISK OF
DROSOPHILA MELANOGASTER[1] **AND THE EPa CELL LINE FROM THE**
EMBRYOS OF *PERIPLANETA AMERICANA*[18,19]

Vitamins	*D. melanogaster*	EPa	Vitamins	*D. melanogaster*	EPa
Thiamine	+	+[a]	Choline	+	+[ab]
Riboflavin	+	+[c]	Ascorbic acid	−	
Pantothenate	+	+[b,c]	Carnitine	−	+[a]
Pyridoxine	+	+[a]	Vitamin A	−	
p-Aminobenzoic acid	−		Vitamin B_{12}	−	+[d]
Folic acid	+	+[c,d]	Vitamin D	−	
Niacinamide	+	+[a]	Vitamin E	−	
Inositol	+	+[b,c]	Vitamin H	−	
Biotin	−		Vitamin K	−	

Note: + = essential; − = nonessential.

[a] Essential for cell growth.
[b] Essential for lipid synthesis.
[c] Essential for long-term survival.
[d] Essential for nucleic acid synthesis.

media are only available for limited cell lines, and, therefore, essentiality of vitamins has not been studied widely. So far, such studies have been made on *Drosophila melanogaster* imaginal disks and on a cell line from American cockroach (*Periplaneta americana*) embryos (Table 7).

In the *P. americana* cell line, folic acid and vitamin B_{12} are said to be necessary for synthesis of nucleic acids and pantothenate; choline and inositol, for synthesis of lipids.[18] It has been shown that four vitamins (calcium pantothenate, folate, riboflavin, and inositol) were required for long-term survival of various cell lines and cell clones from *P. americana*; five others (carnitine, choline, pyridoxine, thiamine and nicotinamide) were essential for their proliferation. The requirement of vitamin B_{12} and ascorbic acid was dependent on the type of cell lines. Biotin and *p*-aminobenzoic acid were not required by any of the *P. americana* cell lines.[19] Grace[20] noted that the addition of ten vitamins to Wyatt's medium was growth-promoting at a concentration of 0.01 μg/ml. However, higher concentrations above 10 μg/ml were rather growth inhibiting. There are several vitamins which are reportedly growth stimulative. Choline, biotin, and inositol stimulated growth of ovarial tissues of the cynthia moth (*Philosamia cynthia*).[21] Increase of choline concentration or pyridoxal caused promotion of growth of yellow fever mosquito (*Aedes aegypti*) cells.[22] Pyridoxal (but neither pyridoxine nor pyridoxamine) brought enhancement of cloning efficiency of *D. melanogaster* cells when its concentration was increased about 1000 times.[23]

VI. LIPIDS

There is some information on the consumption of lipids by insect cells cultured *in vitro*. However, essentiality of lipids for the growth of insect cells has not yet been studied. Most insect cell culture media are used with insect hemolymph or vertebrate sera. Lipids are supposedly supplied from these blood components. Therefore, most culture media are formulated without added lipids.

It is well known that sterols are essential nutrients for insects, because insects cannot synthesize sterols. This is also true for cultured insect cells. Thus, sterols, especially cholesterol, are generally considered essential for the growth of insect cells,[24,25] while vertebrate

Table 8
COMPOSITION OF MM² AND MTCM-1103[30]
MEDIA (MG/100 ml)

Ingredients	MM	MTCM-1103
NaCl	700	700
NaH$_2$PO$_4$	20	20
NaHCO$_3$	12	12
KCl	20	20
CaCl$_2$·2H$_2$O	20	20
MgCl$_2$·6H$_2$O	10	10
Lactalbumin hydrolysate	650	1500
TC-Yeastolate	500	—
Glucose	400	500
Inosine	—	20
Thiamine hydrochloride	—	0.016
Riboflavin	—	0.016
Folic acid	—	0.016
Biotin	—	0.008
Calcium pantothenate	—	0.016
Pyridoxine hydrochloride	—	0.016
p-Aminobenzoic acid	—	0.016
Niacin	—	0.016
i-Inositol	—	0.016
Choline chloride	—	0.016
Fetal bovine serum	0—10 (ml)	—
pH (with KOH)	6.3	6.3

Modified from Mitsuhashi, J., *Appl. Entomol. Zool.*, 17, 575, 1982. With permission.

cells do not require sterols because these cells can synthesize sterols. The addition of cholesterol to the serum-containing media did not improve cell growth,[20,26] although it is certain that insect cells in cultures consume cholesterol.[27,28] The *Antheraea eucalypti* cell line was said to consume 4 to 5% of sterol in the medium during an 8-d culture.[28] In serum-free cultures of *Blattera germanica* embryonic cell lines, cholesterol should be supplied as coarse granules, so that the protective action of the membrane is not lost.[29] However, essentiality of sterols for insect cells *in vitro* may need reconsideration.

Recently, several insect cell lines were able to be cultured continuously with media free of proteins and lipids. An example of such media is MTCM-1103 (Table 8).[30] This fact indicates that at least some insect cell lines such as the cell lines from the common armyworm (*Leucania separata*), the one-striped mosquito (*Aedes albopictus*), the flesh fly (*Sarcophaga peregrina*), *Papilio xuthus,* and *Mamestra brassicae* do not require sterols for their growth. In fact, sterols could not be detected in the cells of *M. brassicae, A. albopictus,* and *S. peregrina* cultured in MTCM-1103 for many generations (Table 9).[31,32] A similar result has been obtained in *Drosophila melanogaster* and *A. albopictus* cell cultures.[33] The *Periplaneta americana* cell line, EPa, has also been reported to be able to multiply in sterol-free medium.[34] From these results, it may be said that, in some insect cell lines, sterols are nonessential for maintaining the life and multiplication of the cell lines.

In general, sterols are considered an important constituent of the biological membrane. How is the membrane of the above-mentioned sterol-free insect cell constructed? This is an interesting question. In vertebrate cells, sterols can be detected in cells cultured in lipid-free media, because they can synthesize sterols. When the cells are cultured in lipid-free media with a sterol synthesis inhibitor, the sterol content of the cells decreases markedly, and in such a medium the cells cannot be cultured continuously. Therefore, in the animal kingdom,

Table 9
STEROL CONTENTS OF SEVERAL INSECT CELL LINES

Cell lines[a]	Dry weight (mg)	Protein content (mg)	Total lipid content (mg)	Sterol content	
				(μg)	(μg/mg protein)
MB-25 (FS)[b]	118	64.8	11.6	ND[c]	<0.00077[d]
MB-25 (WS)	91	50.1	13.3	198	3.95
MB-32 (FS)	98	47.3	14.9	ND	<0.00106[d]
MB-32 (WS)	85	41.1	13.0	120	2.92
MaBr-85 (FS)	128	70.0	—	ND	<0.00071[d]
MaBr-85 (WS)	98	53.6	15.5	467	8.71
AeAl-2 (FS)	82	25.5	14.1	ND	<0.00196[d]
AeAl-2 (WS)	88	36.2	18.8	451	12.45
SaPe-4 (FS)	98	48.7	10.0	ND	<0.00103[d]
SaPe-4 (WS)	107	53.2	14.0	113	2.12

[a] MB-25: NIAS-MB-25 cell line derived from *Mamestra brassicae* ovarioles; MB-32: NIAS-MB-32 cell line derived from *M. brassicae* ovarioles; MaBr-85: NIAS-MaBr-85 cell line derived from *M. brassicae* fat bodies; AeAl-2: NIAS-AeAl-2 cell line derived from *Aedes albopictus* neonate larvae; and SaPe-4: NIH-SaPe-4 cell line derived from *Sarcophaga peregrina* embryos.

[b] (FS) indicates cultured in a sterol-free medium, MTCM-1103; (WS), cultured in a serum-containing medium, MM with 10% fetal bovine serum.

[c] Not detected.

[d] The value was calculated based on the sensitivity of the analytical system and the quantity of the samples, assuming the cells contained sterols.

From Mitsuhashi, J., Nakasone, S., and Horie, Y., *Cell Biol. Int. Rep.*, 7, 1057, 1983. With permission.

the insect cell line is the first instance of cells which are able to grow free from sterols. In prokaryotes, organisms which can multiply without sterols are not rare, and in the plant kingdom, some fungi are reportedly able to multiply under the sterol-free condition.[35]

Some fatty acids have been considered essential for the growth of insect cells. Usually, sufficient amounts of fatty acids are contained in serum which is added to the culture medium. Thus, very few insect cell culture media have been formulated with added fatty acids. The above-mentioned MTCM-1103 medium is also free of fatty acids. It is, therefore, possible to say that fatty acids are not essential for the growth of some insect cell lines such as *M. brassicae*. The insect cells are capable of synthesizing various fatty acids, and the synthesized fatty acids are liberated into the culture medium (Table 10). However, linoleic acid and linolenic acid do not seem to be synthesized by the insect cells.[36]

Vaughn[25] reported that *A. eucalypti* cells consumed linoleic and linolenic acids in the medium quickly, and the final density of the cells was increased by 40% with the addition of lipids containing the above two fatty acids in the medium. He also showed that the triglyceride, trilinolein, was effective in increasing the final cell density. However, he warned that lipids are detrimental to the growth of insect cells if used at high concentrations.

Gottschewski[1] has reported that fat is not necessary for the culture of *D. melanogaster* imaginal disks. Aizawa and Sato[37] added soybean oil to Wyatt's medium, but no improvement was observed in the cultures of *Bombyx mori* tissues. However, the possibility that fatty acid is required by insect cells cannot be denied, because the *A. eucalypti* cell line reportedly produced considerable amounts of free fatty acids when cultured in a medium containing 10% fetal bovine serum.[38]

In addition to fatty acids, insect cells in culture are reported to be able to synthesize various lipids, predominantly phosphatidylethanolamine and phosphatidylcholine.[24]

Table 10

FATTY ACID COMPOSITION OF TOTAL LIPIDS OF CULTURED CELLS

Cell lines	Dry weight	Total lipid (mg/g dry weight)	Total fatty acids (mg/g dry weight)	Individual fatty acids (mg/g dry weight); number in parenthesis is the percentage to the total fatty acids						
				C14:0	C16:0	C16:1	C18:0	C18:1	C18:2	C18:3
NIAS-MB-32[a]	95.0	152.0	30.94	0.296 (1.0)	2.418 (7.8)	8.255 (26.7)	4.204 (13.6)	15.765 (50.9)	ND —	ND —
NIAS-MB-32[b]	85.1	152.8	69.03	0.483 (0.7)	5.661 (8.2)	19.122 (27.7)	9.112 (13.2)	34.033 (49.3)	0.621 (0.9)	ND —
NIAS-AeAl-2[a]	82.0	172.0	53.34	1.707 (3.2)	7.817 (14.7)	12.268 (23.0)	4.610 (8.6)	26.317 (49.3)	ND —	ND —
NIAS-AeAl-2[b]	88.2	213.2	79.84	0.878 (1.1)	12.375 (15.5)	25.388 (31.8)	3.992 (5.0)	36.086 (45.2)	1.038 (1.3)	ND —
NIH-SaPe-4[a]	98.0	102.0	20.54	0.316 (1.5)	4.163 (20.3)	7.551 (36.8)	0.449 (2.2)	8.061 (39.2)	ND —	ND —
NIH-SaPe-4[b]	107.0	130.8	64.40	0.515 (0.8)	13.072 (20.3)	20.800 (32.3)	1.610 (2.5)	28.076 (43.6)	0.322 (0.5)	ND —

[a] MTCM-1103, a serum-free medium containing neither fatty acids nor sterols.
[b] MM medium containing 10% fetal bovine serum.

From Mitsuhashi, J., Nakasone, S., and Horie, Y., *Appl. Entomol. Zool.*, 20, 8, 1985. With permission.

VII. ORGANIC ACIDS

Besides amino acids, some organic acids have been incorporated into many media. The idea of incorporation of such organic acids as malic acid, succinic acid, fumaric acid, and α-ketoglutaric acid is attributed to Wyatt's medium.[7] These organic acids are not essential for the growth of insect cells. According to Wyatt,[7] only malic acid was growth promoting if used singly; however, a marked growth-promoting effect was obtained when they were used together. Citric acid and lactic acid were also used by some investigators; however, they are reportedly detrimental to the growth of a mosquito (*Anopheles stephensi*) cell line.[39]

VIII. PROTEINS AND PEPTIDES

Peptone has been incorporated into some culture media. Although the effect of peptone is not clear, improvement of cultures by adding peptone to media have been reported in the tissue culture of the polyphemus moth (*Antheraea polyphemus*)[40] and *Drosophila melanogaster*.[1,41] It has also been reported that more than 10 mg peptone per milliliter is detrimental to the cellular outgrowth in *D. melanogaster* primary cultures.[41]

Bovine plasma albumin Fraction 5 (BPA) is often incorporated into culture media. BPA seems to be able to execute at least a part of the role of sera or hemolymph. In *Antheraea eucalypti* cell cultures, hemolymph concentration could be reduced from 5 to 1% by adding 1% BPA.[42] Likewise, in mosquito cell cultures (*Aedes aegypti* and *Culex molestus*), BPA could substitute for the whole calf serum.[43]

Tryptose phosphate broth (TPB) has been reported to substitute for insect hemolymph in cultures of the *Spodoptera frugiperda* cell line.[44]

Peptone, BPA, and TPB are not essential for the growth of insect cells, although they sometimes enhance cellular growth. Vail et al.[45] reported that BPA could be removed from the culture medium (TNM-FH) for cabbage looper (*Trichoplusia ni*) cell cultures without deleterious effects.

Fetuin has been reported to improve primary cultures of *D. melanogaster* embryonic cells.[46] Transferrin and globulin are neither essential nor growth promotive for the growth of a *Papilio xuthus* cell line.[4]

Protein hydrolysates have frequently been used for insect tissue culture media. Among them, lactalbumin hydrolysate (LH) is the most popular. According to Wyatt,[7] the addition of hydrolysates of egg albumin, bovine fibrin, or bovine serum albumin to Trager's medium[8] improved the culture of *Bombyx mori* ovarian cells. Marked promotion of cell multiplication was obtained by the addition of LH and TC-yeastolate to the culture medium for the cotton bollworm moth (*Heliothis zea*) cell line.[47] However, the growth of the *Antheraea eucalypti* cell line (erroneously described as a mosquito *Aedes vexans* cell line[48]) was reported to be inhibited by the 1% LH.[49] The hydrolysate of milk casein has about the same effect as LH on the *Sarcophaga peregrina* cell line.[50] These protein hydrolysates consist mostly of free amino acids and are, therefore, used instead of a mixture of individual amino acids. However, they also contain peptides and other unidentified substances. According to Nagasawa et al.,[50] the purification of lactalbumin before enzymatic digestion did not seem to be sufficient, because they found compounds with higher molecular weight than that of α-lactalbumin in LH. Thus, this was considered one of the reasons why the growth-promoting activity of LH varies from one lot to another.

Gottschewski[1] stated that naturally occurring peptides are necessary for tissue culture; in general, glutathione and glycylglycine were important. However, there have been no detailed analyses of peptide requirements of insect cells in cultures.

The MTCM-1103 medium[30] contained LH as the sole chemically undefined component. Therefore, it is considered that LH contains some factors for cell growth promotion. The factors were found to be water soluble and heat stable. They could stand autoclaving for 15 min. However, the factors lost their activities after treatment with 6 N hydrochloric acid for 20 h at 110°C. Purification of the factors by a gel-filtration column of Sephadex® G-50 resulted in dispersion of the activities into two separate fractions. The molecular weight of the substances in Fraction 1 was estimated to be 4000 to 40,000. From the elution patterns by anion-exchange high performance liquid chromatography (HPLC), Fraction 1 showed a basic nature. The molecular weight of the growth factors in Fraction 2 appeared to be small, and they showed affinity to the Sephadex® gel. Fraction 2 seemed to contain many growth factors with different ionic natures, because scattering of activities was observed in either anion- or cation-exchange HPLC.[50]

IX. NUCLEIC ACID-RELATED SUBSTANCES

There is no definite information on the essentiality of nucleic acid-related substances. However, some seem to have stimulating activity for cell growth, and they are incorporated into culture media by some investigators. Grace[20] reported that yeast RNA and thymus nucleic acid had no effects on ovarian cell cultures of *Bombyx mori*. Gottschewski[1] reported that addition of coenzyme A, cocarboxylase, NAD, TPN, and FAD resulted in enhancement of the ratio of mitoses in *Drosophila melanogaster* imaginal disk cultures. However, he did not ascertain the effect of each substance. Wang et al.[51] reported that an *Antheraea eucalypti* cell line utilized thymidine, but not thymine. Most derivatives and analogs of 2-amino-4-hydroxypteridine were more or less toxic to cultures of *Galleria mellonella* ovarian cell cultures. However, 2-amino-4-mercaptopteridine and its C-6,7-dimethyl derivatives were found to stimulate the outgrowth of ovarian cells of *G. mellonella* at 10 μM.[52] Folate and N-10-formylate have also been reported to promote cellular outgrowth in the same culture system.[53]

X. HORMONES

Ecdysone analogs have been reported to have slight stimulation activity on the growth of an *Antheraea eucalypti* cell line[26] and *Drosophila melanogaster* cells.[46] However, they are not necessary for the growth of cells. In *D. melanogaster* cell lines, ecdysteroid hormones have been known to arrest cell growth and to induce morphological, behavioral, and biochemical changes.[54-57]

Dodecyl methyl ether, which has juvenile hormone activity, has been reported to have a stimulating effect on *D. melanogaster* embryonic cells,[46] whereas farnesol did not show any favorable effect on a cell line of *A. eucalypti*.[26]

The effect of brain hormone has scarcely been studied. Bombyxin, which is a silkworm 4-kDa prothoracicotropic hormone, was not growth stimulating on *Sarcophaga peregrina* cells.[50]

A vertebrate hormone, insulin, has been found in *D. melanogaster* larvae,[58] and it was found to have a growth-promoting action on a *D. melanogaster* cell line.[58,59] However, the stimulatory effect of insulin seems to occur only in limited cases. It has no effects on the *S. peregrina* cell line[60] nor on lepidopteran cell lines.[4]

The effects of hormones of either insects or vertebrates have been studied by using serum-containing media. The use of chemically defined media may give clear answers for the growth-promoting effects of these hormones.

XI. INSECT HEMOLYMPH

In the early stage of insect tissue cultures, hemolymph was used frequently as a major component of culture media. However, melanization of hemolymph often deteriorated the cultures. At present, insect hemolymph is seldom used. However, it is still worthwhile to use it as an additive of the culture media.

Since Wyatt[7] used heat-treated hemolymph, this method became standard for the preparation of insect hemolymph for culture media. Grace[3] treated hemolymph of *Antheraea pernyi* at 60°C for 5 min, and then froze it at −20°C. The treated hemolymph can be stored frozen for a long time until use. Before use, it is thawed, the coagulated protein spun down, and the supernatant used.

Factors for stimulation of cell growth in hemolymph seem to be nonspecific. However, there seems to be some compatibility between donor species of hemolymph and the species whose tissues or cells are cultured. The hemolymph prepared from diapausing pupae of *A. pernyi* seems to be applicable to a wide range of species, even to species other than Lepidoptera, whereas those prepared from *Bombyx mori* have rather limited use.[4]

The effects of hemolymph also vary among developmental stages of the source. It is reasonable when one considers that chemical composition of hemolymph fluctuates with the growth of insects. In the gypsy moth, *Lymantria dispar,* growth-promoting ability of hemolymph was two to three times stronger in the one taken from larvae just after the final larval molting or just before pupation, compared to the one taken from larvae at middle stages of the final instar.[7] Frew[61] reported that larval hemolymph was ineffective on the growth of blowfly imaginal disks, while pupal hemolymph was slightly stimulative. On the other hand, Haskell and Sanborn[62] showed that larval hemolymph was better than the pupal type for the gonads and imaginal disks of *Hyalophora cecropia* and, furthermore, that the latter hemolymph caused cytolysis of explants.

However, Williams and Kambysellis[63] showed with the same species that growth-promoting substance (GPS) was absent in the early stages of diapausing pupae, and it appeared when the pupal cuticle was injured or when ecdysone was injected.

The fluctuation in growth-promoting effects of hemolymph seems to be different in different species.

The GPS in insect hemolymph has not yet been clarified. The GPS in *B. mori* hemolymph was dialyzed, stable against acid and alkali treatments, and was not adsorbed to activated charcoal,[37] whereas that in *A. pernyi* hemolymph was not dialyzed.[65,65] The GPS in *H. cecropia* hemolymph was not dialyzable and heat labile.[63]

According to Vaughn and Louloudes,[66] the GPS activity of *B. mori* hemolymph against a *Spodoptera frugiperda* cell line decreased to 42% if lipid was removed. The GPS was water soluble, and its molecular weight was between 1000 and 5000. The reconstituted hemolymph showed 90% of the original activity. Of the extracted lipids, the activity was found in the neutral lipid fraction consisting of triglycerides, diglycerides, and sterols.

XII. VERTEBRATE SERA

The most frequently used serum of vertebrates is fetal bovine serum (FBS), and, in most cases of primary cultures, insect cells cannot grow without it. It is, therefore, considered that FBS contains some GPS for insect cells. For some insect cell cultures, FBS is said to have been immobilized at 56°C for 30 min.[67,68]

The GPS in FBS has not yet been identified. After dialysis, the activity remained in the undialyzed portion, although it decreased to about 80%.[69,70] The GPS was recovered from the regions of albumin, fetuin, and transferrin by fractionation through BioGel® P-300, and its activity disappeared when it was treated with 6 M urea or pronase.[69,70] By means of Cohen's cold ethanol method, the activity was recovered from Fraction 5.[69,70]

Mitsuhashi[71] investigated the nature of GPS in FBS after various treatments. The activity of GPS was assayed by the use of a *Papilio xuthus* cell line. When FBS was heated for 15 min, the activity decreased with a rise in temperature from 60°C and was completely lost at 80°C. After the FBS was boiled for 15 min and its precipitates were removed, it showed inhibitory effect on the cell growth. From the filtration with Diaflo® membranes — SM-50, XM-100A, and XM-300 (Amicon, Lexington, KY) — the activity of GPS in FBS was retained well in the fraction with molecular weight of 300,000; however, a part of the activity was lost. Furthermore, the reconstituted FBS showed less activity than that of the filtration residue which retained activity, suggesting that the filtrate became inhibitory for cell growth irreversibly.

Besides growth-promoting action, FBS possesses the ability to enhance the cellular adhesiveness to glass[72] and to protect cells from lysis.[73] The latter effect was also obtained by the use of an α-globulin fraction from human, bovine, or horse serum,[73] α_1-antitrypsin,[73] α_2-macroglobulin,[73] protein π,[73] or Ficoll.[74] On the other hand, FBS reportedly contained some toxic substance for *Trichoplusia ni*,[75] several leafhopper cells,[75] and *Drosophila melanogaster* cells.[76]

The newborn calf serum also has GPS, but the activity is considerably lower compared to that of FBS.[2] Chicken serum has been used instead of FBS in cultures of some cell lines such as *T. ni*[77] and *Heliothis zea*.[78] Turkey serum instead of FBS supported the growth of the cells from *Spodoptera frugiperda* and *T. ni* for some time.[67] Other vertebrate sera, such as those from human, bovine, calf, horse, lamb, rabbit, guinea pig, and chicken, usually do not show activity of growth promotion on insect cells.[2,77,78]

XIII. TISSUE EXTRACTS

Chicken egg extract (CEE) had been used often in vertebrate tissue cultures. Recently, it has not been used frequently, because many growth factors for vertebrate cells were isolated. For invertebrate tissue cultures, homologous or heterologous tissue extracts do not seem to be essential. However, they sometimes enhance cell growth. Although Gottschewski[1] has reported that the addition of CEE showed no effect on cultures of *Drosophila melanogaster* imaginal disks, improvement of cultures by incorporation of CEE has been reported in the cultures of lepidopteran ovarian cells,[7,79] the imaginal disks of *D. melanogaster*,[80-82] *Calliphora erythrocephala*,[80-82] and *Aedes aegypti*,[83] and ovaries of the yellow mealworm (*Tenebrio molitor*).[84] The CEE in the culture of *T. molitor* could be replaced with glycoprotein.[84]

Chicken egg ultrafiltrate has been used for some culture media, but it is not essential.[45,77]

For extracts of insect tissues, those from eggs, embryos, or ovaries have sometimes been used. Growth stimulation or prolongation of survival has been obtained in the culture of *B. mori* ovarian cells by the incorporation of homologous embryo extract,[7] ovary extract,[20] or *C. erythrocephara* egg extract;[85] in the culture of *D. melanogaster* or *C. erythrocephala* imaginal disks by homologous egg extract;[80,86] in the culture of hemocytes of the variegated cutworm (*Peridroma saucia*) by house fly (*Musca domestica*) egg extract;[87] and in the culture of embryonic cells of the smaller brown planthopper (*Laodelphax striatellus*) by homologous egg extract.[88] As with the other tissue extracts, those from ring glands,[20] prothoracic glands,[20] fat bodies,[89-91] or whole pupae of *D. melanogaster*[80,86] have been reported to be effective.

REFERENCES

1. **Gottschewski, G. H. M.,** Morphologische Untersuchungen an in vitro wacshsenden Augenanlagen von *Drosophila melanogaster, Wilhelm Roux Arch. Entwicklungsmech. Organ.,* 152, 204, 1960.
2. **Mitsuhashi, J. and Maramorosch, K.,** Leafhopper tissue culture: embryonic, nymphal, and imaginal tissues from aseptic insects, *Contrib. Boyce Thompson Inst.,* 22, 435, 1964.
3. **Grace, T. D. C.,** Establishment of four strains of cells from insect tissues grown in vitro, *Nature (London),* 195, 788, 1962.
4. **Mitsuhashi, J.,** unpublished data, 1987.
5. **Ting, K. Y. and Brooks, M. A.,** Sodium:potassium ratios in insect cell culture and the growth of cockroach cells (Blattariae: Blattidae), *Ann. Entomol. Soc. Am.,* 58, 197, 1965.
6. **Krause, G., Krause, J., and Geisler, M.,** Beobachtungen in vitro an der Zellen der larvalen Ovariolenhülle von *Bombyx mori* L. (Lepidoptera) in verschiedenen Kurturmedien, *Z. Zellforsch. Mikrosk. Anat.,* 70, 393, 1966.
7. **Wyatt, S. S.,** Culture in vitro of tissue from the silkworm, *Bombyx mori* L., *J. Gen. Physiol.,* 39, 841, 1956.
8. **Trager, W.,** Cultivation of the virus of grasserie in silkworm tissue cultures, *J. Exp. Med.,* 61, 501, 1935.
9. **Adam, J. R., Goodwin, R. H., Vaughn, J. L., and Piscopo, I.,** X-ray microanalysis of insect viruses, insect tissue culture cells and insect larval tissues and *Bacillus thuringiensis,* presented at the 8th Annu. Meet. Soc. Invertebr. Pathol., Corvallis, 1975, 10.
10. **Weiss, S. A., Kalter, S. S., Vaughn, J. L., and Dougherty, E.,** Effect of nutritional, biological and biophysical parameters on insect cell culture of large scale production, *In Vitro,* 16, 222, 1980.
11. **Grace, T. D. C. and Brzostowski, H. W.,** Analysis of the amino acids and sugars in an insect cell culture medium during cell growth, *J. Insect Physiol.,* 12, 625, 1966.
12. **Mitsuhashi, J.,** Simplification of media and utilization of sugars by insect cells, in *Invertebrate and Fish Tissue Culture,* Kuroda, Y., Kurstak, E., and Maramorosch, K., Eds., Japan Sci. Soc. Press, Springer-Verlag, Tokyo, 1987, 15.
13. **Kimura, S. and Mitsuhashi, J.,** Preliminary notes on glycosidases in two cell lines derived from the ovary of the cabbage armyworm, *Mamestra brassicae* (Lepidoptera, Noctuidae), *Appl. Entomol. Zool.,* 18, 561, 1983.
14. **Mitsuhashi, J.,** Requirements of amino acids by insect cell lines, in *Invertebrate Systems in Vitro,* Kurstak, E., Maramorosch, K., and Dübendorfer, A., Eds., Elsevier/North-Holland, Amsterdam, 1980, 47.
15. **Mitsuhashi, J.,** Determination of essential amino acids for insect cell lines, in *Invertebrate Cell Culture Applications,* Maramorosch, K. and Mitsuhashi, J., Eds., Academic Press, New York, 1982, 9.
16. **Mitsuhashi, J.,** Insect cell line: amino acid utilization and requirements, in *Invertebrate Tissue Culture, Applications in Medicine, Biology and Agriculture,* Kurstak, E. and Maramorosch, K., Eds., Academic Press, New York, 1976, 257.
17. **Mitsuhashi, J.,** Difference in amino acid metabolism among cell lines derived from cabbage armyworm, *Mamestra brassicae* (Lepidoptera: Noctuidae), *Appl. Entomol. Zool.,* 22, 533, 1987.
18. **Landureau, J. C.,** Études des exigences d'une lignée de cellules d'insectes (souche EPa). II. Vitamines hydrosolubles, *Exp. Cell Res.,* 54, 399, 1969.
19. **Becker, J. and Landureau, J. C.,** Specific vitamin requirements of insect cell lines (*P. americana*) according to their tissue origin and in vitro conditions, *In Vitro,* 17, 471, 1981.
20. **Grace, T. D. C.,** Effects of various substances on growth of silkworm tissues in vitro, *Aust. J. Biol. Sci.,* 11, 407, 1958.
21. **Sanborn, R. C. and Haskell, J. A.,** Chemical requirements for growth of insect tissues in vitro, in *Proc. 12th Int. Congr. Entomol.,* Vienna, 1960, BIII 237.
22. **Nagle, S. C.,** Improved growth of mammalian and insect cells in media containing increased levels of choline, *Appl. Microbiol.,* 17, 318, 1969.
23. **Wyss, C.,** Cloning of *Drosophila* cells: effects of vitamins and yeast extract components, *Somatic Cell Genet.,* 5, 23, 1979.
24. **Cohen, E. and Gilbert, L. I.,** Lipid synthesis and RNA thermolability in the *Trichoplusia ni* cell line, *Insect Biochem.,* 5, 671, 1975.
25. **Vaughn, J. L.,** Insect cell nutrition: emphasis on sterols and fatty acids, *In Vitro,* 9, 122, 1973.
26. **Mitsuhashi, J. and Grace, T. D. C.,** The effects of insect hormones on the multiplication rates of cultured insect cells in vitro, *Appl. Entomol. Zool.,* 5, 182, 1970.
27. **Gilby, A. R. and McKellar, J. W.,** The utilization of sterols and other lipids by insect cells grown in vitro, *J. Insect Physiol.,* 20, 2219, 1974.
28. **Vaughn, J. L., Louloudes, S. J., and Dougherty, K.,** The uptake of free and serum-bound sterols by insect cells in vitro, *Curr. Top. Microbiol. Immunol.,* 55, 92, 1971.
29. **Brooks, M. A. and Tsang, K. R.,** Replacement of serum with cholesterol and lipids for cell lines of *Blattera germanica, In Vitro,* 16, 222, 1980.

30. **Mitsuhashi, J.,** Continuous cultures of insect cell lines in media free of sera, *Appl. Entomol. Zool.,* 17, 575, 1982.
31. **Mitsuhashi, J., Nakasone, S., and Horie, Y.,** Sterol-free eukaryotic cells from continuous cell lines of insects, *Cell Biol. Int. Rep.,* 7, 1057, 1983.
32. **Mitsuhashi, J.,** Serum-free culture and sterol requirements of insect cell lines, presented at the 6th Int. Conf. Invertebr. Tissue Culture, St. Augustine, FL, June 5 to 10, 1983, 44.
33. **Silberkang, M., Havel, C. M., Friend, D. S., McCarthy, B. J., and Watson, J. A.,** Isoprene synthesis in isolated embryonic *Drosophila* cells. I. Sterol-deficient eukaryotic cells, *J. Biol. Chem.,* 254, 8503, 1983.
34. **Landureau, J. C. and Jollès, P.,** Étude des exigenses d'une lignée de cellules d'insectes (souche EPa). I. Acides amines, *Exp. Cell Res.,* 54, 391, 1969.
35. **Elliot, C. G., Hendric, M. E., Knights, B. A., and Parker, W.,** A steroid growth factor requirement in a fungus, *Nature (London),* 4943, 427, 1964.
36. **Mitsuhashi, J., Nakasone, S., and Horie, Y.,** Total fatty acids of some insect cell lines, *Appl. Entomol. Zool.,* 20, 8, 1985.
37. **Aizawa, K. and Sato, F.,** Culture de tissus de ver à soie, *Bombyx mori,* dans un milieu sans hemolymphe, *Ann. Epiphyt.,* 14, 125, 1963.
38. **Jenkin, H., Towsend, H. T., Makino, S., and Yang, T. K.,** Comparative lipid analysis of *Aedes aegypti* and monkey kidney cells (MK-2) cultivated in vitro, *Curr. Top. Microbiol. Immunol.,* 55, 97, 1971.
39. **Schneider, I.,** Establishment of three diploid cell lines of *Anopheles stephensi* (Diptera: Culicidae), *J. Cell Biol.,* 42, 603, 1969.
40. **Loeb, M. J. and Schneiderman, H. A.,** Prolonged survival of insect tissues in vitro, *Ann. Entomol. Soc. Am.,* 49, 493, 1956.
41. **Kuroda, Y.,** Growth stimulating effect of peptone on *Drosophila* ovarian cells in culture, *Annu. Rep. Natl. Inst. Genet.,* No. 21, 42, 1971.
42. **Grace, T. D. C.,** Cultured cells in virus research, in *Viruses and Invertebrates,* Gibs, A. J., Ed., North-Holland, Amsterdam, 1973, 321.
43. **Kitamura, S., Imai, T., and Grace, T. D. C.,** Adaptation of two mosquito cell lines to medium free of calf serum, *J. Med. Entomol.,* 10, 488, 1973.
44. **Gardiner, G. R. and Stockdale, H.,** Two tissue culture media for production of lepidopteran cells and nuclear polyhedrosis viruses, *J. Invertebr. Pathol.,* 25, 363, 1975.
45. **Vail, P. V., Joy, D. L., and Romine, C. L.,** Replication of the *Autographa californica* nuclear polyhedrosis virus in insect cell lines grown in modified media, *J. Invertebr. Pathol.,* 28, 263, 1976.
46. **Kuroda, Y.,** An attempt to obtain long-term culture cells from *Drosophila melanogaster, Annu. Rep. Natl. Inst. Genet.,* No. 20, 28, 1970.
47. **Shapiro, M. and Ignoffo, C. M.,** Growth of *Heliothis zea* (Lepidoptera: Noctuidae) cells adapted to various culture media, *Ann. Entomol. Soc. Am.,* 66, 270, 1973.
48. **Greene, A. E. and Charney, J.,** Characterization and identification of insect cell cultures, *Curr. Top. Microbiol. Immunol.,* 55, 51, 1971.
49. **Sweet, B. H. and McHale, J. S.,** Characterization of cell line derived from *Culiseta inornata* and *Aedes vexans* mosquito, *Exp. Cell Res.,* 61, 51, 1970.
50. **Nagasawa, H., Mitsuhashi, J., and Suzuki, A.,** Growth factors in the lactalbumin hydrolysate to the cells of the fleshfly, *Sarcophaga peregrina,* in *Invertebrate and Fish Tissue Culture,* Kuroda, Y., Kurstak, E., and Maramorosch, K., Eds., Japan Sci. Soc. Press, Springer-Verlag, Tokyo, 1987, 29.
51. **Wang, G. M., Matsumura, F., and Boush, G. M.,** Induction in vitro of protein in the cultured tissue cells of *Aedes aegypti, J. Insect Physiol.,* 16, 1283, 1970.
52. **Saska, J., Brzelakowska-Sztabert, B., and Zielinska, Z. M.,** Sensitivity of insect ovarian tissues to various pteridines as tested in tissue culture, *J. Insect Physiol.,* 18, 1733, 1972.
53. **Zielinska, Z. M. and Saska, J.,** Effect of folate and some of its analogues on insect ovaries in vitro, in *Proc. 3rd Int. Colloq. Invertebr. Tissue Culture,* Řeháček, J., Blaškovič, D., and Hink, W. F., Eds., Publ. House Acad. Sci., Bratislava, Czechoslovakia, 1973, 167.
54. **Courgeon, A. M.,** Effect of α and β-ecdysone on in vitro cell multiplication in *Drosophila melanogaster, Nature New Biol.,* 238, 250, 1972.
55. **Courgeon, A. M.,** Action of insect hormones at the cellular level. Morphological changes of a diploid cell line of *Drosophila melanogaster,* treated with ecdysone and several analogues in vitro, *Exp. Cell Res.,* 74, 327, 1972.
56. **Cherbas, P., Cherbas, L., and Williams, C. M.,** Induction of acetylcholinesterase activity by β-ecdysone in a *Drosophila* cell line, *Science,* 197, 275, 1977.
57. **Cherbas, L., Yonge, C. D., Cherbas, P., and Williams, C. M.,** The morphological response of Kc-H cells to ecdysteroids: hormonal specificity, *Roux Arch.,* 189, 1, 1980.
58. **Seecof, R. L. and Dewhurst, S.,** Insulin is a *Drosophila* hormone and acts to enhance the differentiation of embryonic *Drosophila* cells, *Cell Differ.,* 3, 63, 1974.

59. **Mosna, G. and Barigozzi, C.**, Stimulation of growth by insulin in *Drosophila* embryonic cells in vitro, *Experientia*, 32, 855, 1976.
60. **Takahashi, M., Mitsuhashi, J., and Ohtaki, T.**, Establishment of a cell line from embryonic tissues of the fleshfly, *Sarcophaga peregrina* (Insecta: Diptera), *Dev. Growth Differ.*, 22, 11, 1980.
61. **Frew, J. G. H.**, A technique for the cultivation of insect tissues, *J. Exp. Biol.*, 6, 1, 1928.
62. **Haskell, J. and Sanborn, R. C.**, The culture of insect cells in vitro, *Anat. Rec.*, 132, 452, 1958.
63. **Williams, C. M. and Kambysellis, M. P.**, In vitro action of ecdysone, *Proc. Natl. Acad. Sci. U.S.A.*, 63, 231, 1969.
64. **Suitor, E. C., Jr.**, Arthropod tissue culture: a brief outline of its development and description of several of its applications, *Lect. Rev. Ser. U.S. Nav. Med. Res. Unit No. 2*, NAMURU-2LR-023, 1, 1966.
65. **Hsu, S. H., Liu, H. H., and Suitor, E. C., Jr.**, Further description of a subline of Grace's mosquito (*Aedes aegypti* L.) cells adapted to hemolymph-free medium, *Mosq. News*, 29, 439, 1969.
66. **Vaughn, J. L. and Louloudes, S. J.**, Isolation of two growth promoting fractions from insect hemolymph, *In Vitro*, 14, 351, 1978.
67. **Goodwin, R. H.**, Insect cell culture: improved media and methods for initiating attached cell lines from the Lepidoptera, *In Vitro*, 11, 369, 1975.
68. **Martinez-Lopez, G. and Black, L. M.**, Development of a new medium for the culture of agallian leafhopper cells, *In Vitro*, 13, 777, 1977.
69. **Kuno, G.**, Studies of Growth-Promoting Proteins in Fetal Bovine Serum Using *Aedes aegypti* Cells Cultured in Vitro, Ph.D. thesis, Ohio State University, Columbus, 1970.
70. **Kuno, G., Hink, W. F., and Briggs, J. D.**, Growth-promoting serum proteins for *Aedes aegypti* cells cultured in vitro, *J. Insect Physiol.*, 17, 1865, 1971.
71. **Mitsuhashi, J.**, Studies on elucidation and application of physiologically active substances by means of insect tissue culture, in *Development of Productive Activity of Organisms* (in Japanese), Res. Rep. Educ. Comm., Ministry of Education, Japan, 1980, 364.
72. **Peters, D. and Black, L. M.**, Techniques for the cultivation of cells of the aphid, *Acyrthosiphon pisum* in primary cultures, *Tagungsber. Dtsch. Akad. Landwirtschaftswiss. Berlin*, 115, 129, 1971.
73. **Landureau, J. C. and Steinbach, M.**, In vitro cell protective effects by certain antiproteases of human serum, *Z. Naturforsch.*, 25b, 231, 1970.
74. **Landureau, J. C.**, Role biologique de la vitamine B_{12}: analyse de l'exigence vitaminique stricte d'une lignée cellulaire d'insectes in vitro, *C. R. Acad. Sci.*, D-270, 3288, 1970.
75. **McIntosh, A. H., Evers, D., and Shamy, R.**, A toxic substance in fetal bovine serum, *In Vitro*, 12, 302, 1976.
76. **Davis, K. T. and Shearn, A.**, In vitro growth of imaginal disks from *Drosophila melanogaster*, *Science*, 196, 438, 1977.
77. **Hink, W. F., Strauss, E., and Mears, J. L.**, Effects of media constituents on growth of insect cells in stationary and suspension cultures, *In Vitro*, 9, 371, 1974.
78. **Ignoffo, C. M., Shapiro, M., and Dunkel, D.**, Effects of insect, mammalian, and avian sera on in vitro growth of *Heliothis zea* cells, *Ann. Entomol. Soc. Am.*, 66, 170, 1973.
79. **Vago, C. and Chastang, S.**, Cultures de tissus d' insectes a l 'aide de serum de mammiferes, *Entomophaga*, 7, 175, 1962.
80. **Demal, J.**, Culture in vitro d'ebauches imaginales de Diptères, *Ann. Sci. Nat. Zool. Biol. Anim.*, 18, 155, 1956.
81. **Demal, J. and Leloup, A.-M.**, Essai de culture in vitro d'organes d'insectes, *Ann. Épiphyt.*, 14 (Ser. III), 91, 1963.
82. **Demal, J.**, Culture in vitro d'ebauches imaginales de Diptères, *Ann. Sci. Nat. Zool. Biol. Anim.*, 18, 155, 1965.
83. **Peleg, J. and Trager, W.**, Cultivation of insect tissues in vitro and their application to the study of arthropod-borne viruses, *Am. J. Trop. Med. Hyg.*, 12, 820, 1963.
84. **Laverdure, A.-M.**, Les apports de la technique des cultures in vitro dans l'etude de la vitellogenèse chez *Tenebrio molitor* (Coleoptère), *C. R. Acad. Sci.*, D-278, 1511, 1974.
85. **Krause, G.**, Zum Verhalten explantierter Larvalgewebe und Embryonalanlagen des Seidenspinners, *Bombyx mori* L. in hängenden Tropfen, *Verh. Dtsch. Zool. Ges. (Vienna)*, 190, 1962.
86. **Demal, J.**, Problemes concernant la morphogenèse in vitro chez les insectes, *Bull. Soc. Zool. Fr.*, 86, 522, 1961.
87. **Martignioni, M. E. and Scallion, R. J.**, Preparation and uses of insect hemocyte monolayers in vitro, *Biol. Bull.*, 121, 507, 1961.
88. **Mitsuhashi, J.**, Preliminary report on the primary cultures of smaller brown planthopper cells in vitro (Hemiptera, Delphacidae), *Appl. Entomol. Zool.*, 4, 151, 1969.
89. **Laverdure, A.-M.**, Culture in vitro des ovaires de *Tenebrio molitor* (Coleoptère). Importance de la composition du milieu sur la survie, la croissance et la vitellogenèse, *C. R. Acad. Sci.*, D-265, 505, 1967.
90. **Lender, Th. and Laverdure, A.-M.**, Culture in vitro des ovaires de *Tenebrio molitor* (Coleoptère). Croissance et vitellogenèse, *C. R. Acad. Sci.*, D-265, 451, 1967.

91. **Lender, Th. and Laverdure, A.-M.,** Cultures organotypiques des ovaires de *Tenebrio molitor* (Coleoptère) < in vitro >. Survie, croissance et vitellogenèse, in *Proc. 2nd Int. Colloq. Invertebr. Tissue Culture,* Barigozzi, C., Ed., Istituto Lombardo Accademia di Scienze e Lettere, Milan, 1968, 138.
92. **Kurtti, T. J., Chaudhary, S. P. S., and Brooks, M. A.,** Influence of physical factors on the growth of insect cells in vitro. II. Sodium and potassium as osmotic pressure regulators of moth cell growth, *In Vitro,* 11, 274, 1975.
93. **Hirumi, H. and Maramorosch, K.,** The in vitro cultivation of embryonic leafhopper tissues, *Exp. Cell Res.,* 36, 625, 1964.
94. **McIntosh, A. H., Maramorosch, K., and Rechtoris, C.,** Adaptation of an insect cell line (*Agallia constricta*) in a mammalian cell culture medium, *In Vitro,* 8, 375, 1973.
95. **Wyss, C. and Bachmann, G.,** Influence of amino acids, mammalian serum, and osmotic pressure on the proliferation of *Drosophila* cell lines, *J. Insect Physiol.,* 22, 1581, 1976.
96. **Adam, G. and Sander, E.,** Isolation and culture of aphid cells for the assay of insect-transmitted plant viruses, *Virology,* 70, 502, 1976.
97. **Eppler, A., Adam, G., and Sander, E.,** Nutritional requirements and growth of two cell-lines from the aphid parasite, *Ephedrus plagiator* under different culture conditions, in *Invertebrate Systems in Vitro,* Kurstak, E., Maramorosch, K., and Dübendorfer, A., Eds., Elsevier/North-Holland, New York, 1980, 59.
98. **Weiss, S. A., Smith, G. C., Kalter, S. S., and Vaughn, J. L.,** Improved method for the production of insect cell cultures in large volume, *In Vitro,* 17, 495, 1981.
99. **Sohi, S. S.,** The effect of pH and osmotic pressure on the growth and survival of three Lepidopteran cell lines in *Invertebrate Systems in Vitro,* Kurstak, E., Maramorsch, K., and Dübendorfer, A., Eds., Elsevier/North-Holland, New York, 1980, 35.
100. **Stockdale, H. and Gardiner, G. R.,** Utilization of some sugars by a line of *Trichoplusia ni* cells, in *Invertebrate Tissue Culture: Applications in Medicine, Biology and Agriculture,* Kurstak, E. and Maramorosch, K., Eds., Academic Press, New York, 1976, 267.
101. **Berridge, M. J.,** Metabolic pathways of isolated Malpighian tubules of the blowfly functioning in an artificial medium, *J. Insect Physiol.,* 12, 1523, 1966.
102. **Mitsuhashi, J.,** Utilization and requirements of amino acids by a cell line derived from the butterfly, *Papilio xuthus, J. Insect Physiol.,* 22, 397, 1976.
103. **Mitsuhashi, J.,** Establishment and characterization of continuous cell lines from pupal ovaries of the cabbage armyworm, *Mamestra brassicae* (Lepidoptera, Noctuidae), Dev. Growth Differ., 19, 337, 1977.
104. **Mitsuhashi, J.,** Establishment and some characteristics of a continuous cell line derived from fat bodies of the cabbage armyworm (Lepidoptera, Noctuidae), *Dev. Growth Differ.,* 23, 63, 1981.
105. **Mitsuhashi, J.,** Amino acid requirements of some continuous cell lines of insects, *Appl. Entomol. Zool.,* 13, 170, 1978.

Chapter 2

ADVANCES IN THE DEFINITION OF CULTURE MEDIA FOR MOSQUITO CELLS

T. J. Kurtti and U. G. Munderloh

TABLE OF CONTENTS

I. INTRODUCTION

Mosquitoes are a problem in temperate and tropical zones of the world. The adult female requires vertebrate blood for her eggs to develop and seeks a host soon after emergence. During the blood meal, a variety of human and animal pathogens may be exchanged between host and mosquito, including viruses, bacteria, protozoa, and filarial worms. Unfortunately, the relationship between vector and pathogen is in most cases poorly understood because of a lack of suitable *in vitro* culture systems. We use mosquito cells to study the metabolic characteristics of these invertebrates at the cellular level and their relationship with mosquito-borne protozoan parasites. We are working mainly with two cell lines: one, ASE-1V, isolated by us from a vector of malaria, *Anopheles stephensi*;[1] the other, TAE-12V, isolated from a nonbiting mosquito, *Toxorhynchites amboinensis*.[2]

Two media used to culture mosquito cells are Eagle's MEM and Leibovitz's L-15. We routinely employ L-15, which has some unusual features. Its buffer system is based on phosphates and amino acids, not on bicarbonate and CO_2. Two other prominent characteristics are the replacement of glucose by galactose and pyruvate, and an unusually high concentration of amino acids. As a benefit, cultures may be incubated in open gas exchange with the atmosphere in the absence of CO_2.[3]

Malarial parasites depend on their host for a supply of exogenous purines, primarily hypoxanthine (H), as they cannot synthesize them *de novo*.[4] Whole *Anopheles albimanus* incorporated adenine (A),[5] and the salivary glands of a tick vector (*Rhipicephalus appendiculatus*) also incorporated A.[6] Because both midgut and salivary glands are sites of major parasite development, their purine-pyrimidine utilization patterns might regulate, or otherwise influence, parasite development. Conversely, the nucleic acid metabolism of mosquito cells in culture could be expected to play an important role in their use as substrates for parasite development *in vitro*.

One drug that can be used to dissect the purine-pyrimidine metabolism is aminopterin. It blocks the *de novo* synthesis of purines and pyrimidines by inhibiting the enzyme dihydrofolate reductase. Purines and pyrimidines, present in supplements (e.g., yeastolate or tryptose phosphate broth [TPB]) to media, can greatly confuse such studies, because they permit cells to complete nucleic acid synthesis via salvage pathways, even in the presence of the aminopterin.

II. MATERIALS AND METHODS

A. Cell Lines

Mosquito cell lines derived from two different genera were used. These were TAE-12V from *T. amboinensis*,[2] and ASE-1V from *A. stephensi*[1] isolated from embryonic material according to the methods of Varma et al.[7] They were incubated at 30°C. Cell growth was estimated by measuring the amount of protein[8] in three replicate 2-ml culture tubes (Nunc, Denmark) as outlined by Kurtti et al.[9] Giemsa-stained chromosome spreads were prepared from cells lysed by hypo-osmotic shock and were fixed with a mixture of two parts acetic acid to three parts of methanol.

B. Culture Media

Leibovitz's L-15 medium[3] (GIBCO, Grand Island, NY) was used for the isolation and initial maintenance of both cell lines. It was supplemented with 10% TPB (Difco, Detroit, MI) and 20% fetal bovine serum (FBS) (GIBCO). The final pH was adjusted with HCl to 7.0. Sera were preferentially used that were not heat inactivated, if possible, and always pretested on each line before purchase of a particular lot.

During our studies on nutritional requirements of mosquito cells, L-15B[10] was developed

Table 1
ADDITIVES TO L-15[3] IN THE FORMULATION OF L-15B[10]

	mg/l		mg/l
L-Aspartic acid	299	$FeSO_4 \cdot 7H_2O$	0.500
L-Proline	300	$CoCl_2 \cdot 6H_2O$	0.002
L-Glutamic acid	500	$CuSO_4 \cdot 5H_2O$	0.002
L-Glutamine	292	$MnSO_4 \cdot H_2O$	0.016
α-Ketoglutaric acid	299	$ZnSo_4 \cdot 7H_2O$	0.020
D-Glucose	2239	$NaMoO_4 \cdot 2H_2O$	0.002
Glutathione (reduced)	10	Na_2SeO_3	0.002
Ascorbic acid	10		
p-Aminobenzoic acid	1		
Vitamin B_{12}	0.5		
d-Biotin	0.1		

(Table 1). It permits cultivation of mosquito cells in the absence of TPB and with reduced serum concentrations. This medium was adjusted to the desired pH using NaOH and was supplemented with 1 to 5% FBS.

C. Purines and Pyrimidines

The following compounds were purchased from Sigma (St. Louis, MO): adenine (A), adenosine (AR), hypoxanthine (H), inosine (HR), guanine (G), guanosine (GR), cytosine (C), uridine (UR), and thymidine (TdR). They are prepared as an aqueous 10 mM stock, filter sterilized, and diluted as appropriate before use. Various combinations of purines and pyrimidines were tested with respect to their ability to replace TPB and to salvage cells made auxotrophic by addition of aminopterin (purchased from Sigma). The influence of these compounds on cell proliferation was assessed by a growth assay as outlined above (Section IIA).

III. RESULTS AND DISCUSSION

A. Cell Lines

The two mosquito lines became established in culture rapidly and could be subcultured on a regular basis within 4 months.[11] By selectively transferring only unattached cells, we developed lines that grew in suspension as fluid-filled multicellular spheres of epithelial-like cells. Once a week they were pipetted repeatedly to disrupt the vesicles, and diluted 1:20. Because they stemmed from cultures seeded with embryonic fragments, their tissue of origin was unknown.

Chromosome spreads can distinguish between mosquito cells and other invertebrate genera on the basis of chromosome numbers and their morphology. Both the ASE-IV and TAE-12V line were predominantly diploid with six chromosomes in about 90% of the cells. However, this method does not separate congeners.

B. Culture Media

In the isolation of our cell lines, L-15 was heavily supplemented with 20% FBS and 10% TPB. Such a highly undefined medium made it impossible to analyze all but the most basic needs of the cells. During our studies on the nutritional requirements of the two lines, we found it necessary to reduce, or even eliminate, undefined supplements by replacing them, in part, with defined components not present in the original formula. This modified L-15, called L-15B (Table 1),[10] included additional amino acids found beneficial for *Aedes al-*

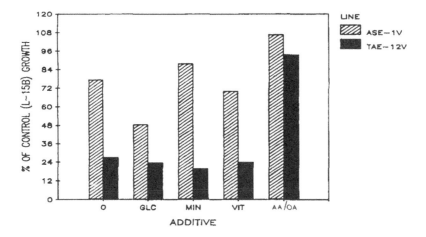

FIGURE 1.   Effects of additives to Leibovitz's L-15 with 2% FBS on the growth of mosquito cells. The substances were added to the medium in the concentration given in Table 1. Cell yield after 10 d was compared to cultures grown in L-15B with 2% FBS. 0 = no additives; GLC = glucose; MIN = minerals (Table 1), including ascorbic acid and reduced glutathione; VIT = biotin and vitamin B_{12}; AA/OA = amino acids plus α-ketoglutaric acid.

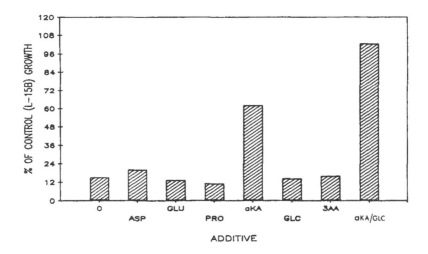

FIGURE 2.   Effects of additives to Leibovitz's L-15 with 2% FBS on the growth of TAE-12V cells. The substances were added to the medium in the concentration given in Table 1. Cell yield after 10 d was compared to cultures grown in L-15B with 2% FBS. 0 = no additives; ASP = aspartic acid; GLU = glutamic acid; PRO = proline; αKA $\approx$ α-ketoglutaric acid; GLC = glucose; 3AA = aspartic acid, glutamic acid, and proline; αKA/GLC = α-ketoglutaric acid plus glucose.

bopictus cells by Mitsuhashi,[12] reduced glutathione, glucose, α-ketoglutaric acid, vitamins, and trace minerals.

To evaluate which of the modifications of L-15B were crucial to the culture of mosquito cells in media with 2% serum, we divided the additives into groups of related compounds. The major benefit was derived from the amino acid to organic acid fraction (Figure 1). Upon closer examination, we found that this was not due to any of the amino acids but rather to α-ketoglutaric acid. Its beneficial effect was further enhanced to 100% of the control (i.e., cells grown in complete L-15B) by the joint use of α-ketoglutaric acid and glucose (Figure 2).

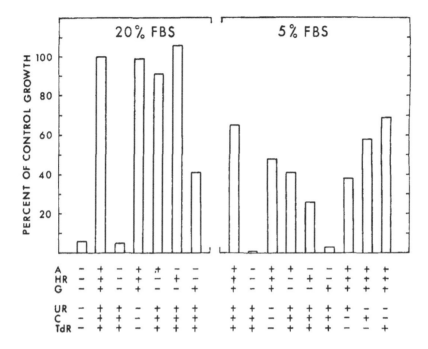

FIGURE 3.   Effect of various combinations of purines and pyrimidines listed below (all
10^{-4} *M*) on TAE-12V cells grown in Leibovitz's L-15 without TPB and with the FBS
concentration shown. Cell yield after 10 d was compared to cultures supplied with 10% TPB.
A = adenine; HR = inosine; G = guanine; UR = uridine; C = cytosine; TdR = thymidine.

C. Purines and Pyrimidines

Immediately after deleting TPB from the medium, the growth of mosquito cells was only
7% of that for the control. However, it could be restored by addition of purines and pyrimidines. In the study summarized in Figure 3, the TAE-12V line was cultured in L-15
supplemented with 20 or 5% FBS in the absence of TPB. Instead, the purine-pyrimidine
combinations shown at the bottom of the graph were added to the test media. When a cocktail
of three purines and three pyrimidines was added, growth was restored to normal levels. In
the presence of 20% serum, this growth stimulation was primarily due to the purines A or
HR, but with 5% FBS, an additional benefit from TdR was realized. Subsequently, we
maintained our mosquito lines in media in which TPB was replaced by the purines A or HR
and the pyrimidine TdR.

During early transfers away from TPB, we noticed that the mosquito cell vesicles accumulated aggregates of large crystals in the lumen (Figure 4). The identity of the crystals
was not determined, and this property was lost after several transfers.

At the same time, we reduced the level of FBS from 20 to 2% over a period of 10 to 20
transfers. When the cells had adapted to growth in medium supplemented with 2% FBS and
0.1 m*M* HR and TdR each, their purine and pyrimidine requirements were again examined
(Figure 5). The main benefit now came from the addition of A alone, and, subsequently,
TdR was deleted. After four to five further transfers, A could be omitted as well. Both the
ASE-IV and TAE-12V lines are currently maintained in L-15B supplemented with 1% FBS
alone.

In the two mosquito cell lines, both *de novo* and salvage pathways of nucleotide metabolism
were functional. When we added aminopterin at a concentration of 1 μ*M* to ASE-IV or
TAE-12V cells in L-15B with 2% FBS, growth was completely blocked. But, as expected
from earlier findings, the purines HR and A — but not AR, G, GR, or H — and the

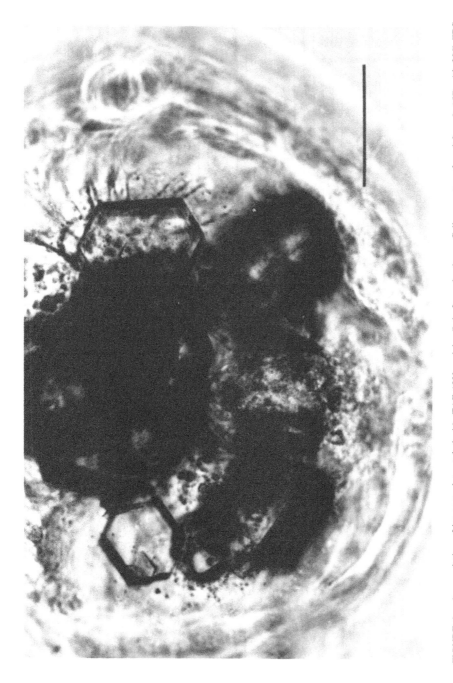

FIGURE 4. Accumulation of hexagonal crystals inside TAE-12V vesicles 7 d after subculture. Cells were transferred from L-15B with 20% FBS and 10% TPB to L-15B with 20% FBS. Bar represents 0.2 mm.

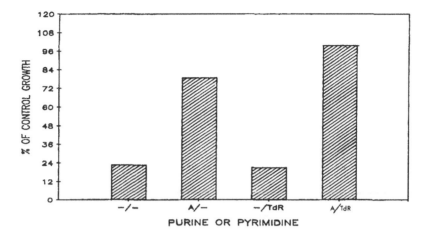

FIGURE 5.   Effect of adenine (A) and thymidine (TdR) at a concentration of 10^{-4} M each on the growth of TAE-12V cells adapted to L-15B supplemented with 2% FBS, A, and TdR (10^{-4} M each). Cell yield after 10 d was compared to cultures grown in the presence of both A and TdR.

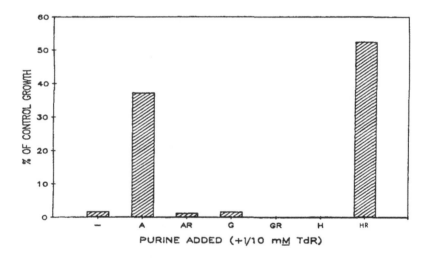

FIGURE 6.   Ability of various purines at a concentration of 10^{-4} M, and in the presence of 0.1 mM thymidine (TdR), to rescue ASE-IV cells made auxotrophic by 10^{-6} M aminopterin in L-15B with 2% FBS. Cell yield was determined after 10 d. A = adenine; AR = adenosine; G = guanine; GR = guanosine; H = hypoxanthine; HR = inosine.

pyrimidine TdR could rescue cells made auxotrophic by aminopterin. With either TdR or HR alone, growth was not restored. The cells needed both. However, with 0.1 mM of each compound, the cell yield was only 53% of the control (Figure 6). This observation led us to explore the influence of concentration and molar ratios of HR and TdR on the cells (Figure 7). One mM of HR plus 0.01 mM of TdR restored to normal the growth of aminopterin-treated ASE-IV cells. Combinations of other purines and pyrimidines were also evaluated but were not as effective.

D. Conclusions

Arthropod cells in culture either synthesize purine and pyrimidine nucleotides *de novo* or salvage bases from the medium. Wilkie et al.[13] added H to a chemically defined medium

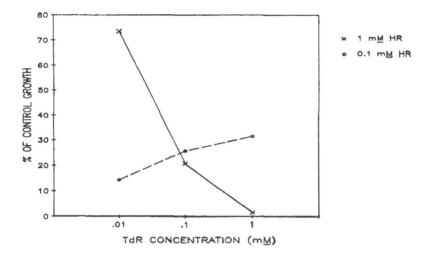

FIGURE 7.   Influence of pyrimidine (TdR = thymidine) to purine (HR = inosine) balance on ASE-IV cells grown in L-15B with 2% FBS and 10^{-6} *M* aminopterin. Cell yield was determined after 10 d.

for mosquito cells, but our results, along with those of Malinoski and Stollar[14] indicated that mosquito cells are unable to utilize H. Wyss[15] found that A, Ar, and HR were salvaged by *Drosophila* cells, and Mitsuhashi[16] incorporated HR into his medium for an *Aedes albopictus* line. We demonstrated that in mosquito cell culture, supplements such as TPB can be replaced by purines and pyrimidines if presented appropriately balanced. When these are supplied, the cells rely primarily on salvage pathways, and enzymes of *de novo* synthesis pathways are repressed. In the absence of these compounds, the latter enzymes are active, but it is important that metabolic precursors of the biosynthetic pathways be present, namely aspartic acid, glutamine, glycine, and folic acid. Even in the presence of precursors, the cells go through long adaptation periods before they acquire the capability of *de novo* synthesis of purines and pyrimidines. The large energy expenditure, in terms of ATP consumption, required to synthesize purines and pyrimidines *de novo* as compared to their utilization via salvage pathways may explain the beneficial effect of these compounds on cells capable of their synthesis. This may also explain the growth-stimulatory effect of the tricarboxylic acid cycle intermediate, α-ketoglutaric acid.

The intracellular pool size and composition of metabolic intermediates in *A. albopictus* cells are influenced by the presence of purines in the medium.[17] Alterations in their natural balance affect the control mechanisms that regulate *de novo* purine and pyrimidine synthesis. Nothing is known about the critical regulatory steps for these pathways in mosquito cells, nor about which molecules play important regulatory roles.[18] This area of research is in need of further investigation. Unraveling the intermediary metabolism of invertebrate cells is likely to be pivotal to our understanding of the interaction between parasitic protozoa and their mosquito vectors.

ACKNOWLEDGMENTS

This is paper No. 15,260, Scientific Journal Series, Minnesota Agricultural Experiment Station. Part of the work presented was carried out while the authors were affiliated with the Department of Entomology and Economic Zoology of Rutgers University, NJ. The work was supported by state funds and a grant from the National Institutes of Health, AI 18345.

REFERENCES

1. **Kurtti, T. J. and Munderloh, U. G.,** unpublished data, 1987.
2. **Munderloh, U. G., Kurtti, T. J., and Maramorosch, K.,** *Anopheles stephensi* and *Toxorhynchites amboinensis:* aseptic rearing of mosquito larvae on cultured cells, *J. Parasitol,* 68, 1085, 1982.
3. **Leibovitz, A.,** The growth and maintenance of tissue-cell cultures in free gas exchange with the atmosphere, *Am. J. Hyg.,* 78, 173, 1963.
4. **Ifediba, T. and Vanderberg, J. P.,** Complete *in vitro* maturation of *Plasmodium falciparum* gametocytes, *Nature (London),* 294, 364, 1981.
5. **Miller, S.,** Utilization and interconversion of purines and ribonucleosides in the mosquito *Aedes albimanus* Weidemann, *Comp. Biochem. Physiol.,* 66B, 517, 1980.
6. **Irvin, A. D., Boarer, C. D. H., Kurtti, T. J., and Ocama, J. G. R.,** The incorporation of radio-labelled nucleic acid precursors by *Theileria parva* in bovine blood and salivary glands of *Rhipicephalus appendiculatus* ticks, *Int. J. Parasitol.,* 11, 451, 1981.
7. **Varma, M. G. R., Pudney, M., and Leake, C. J.,** Methods in mosquito cell culture, in *Practical Tissue Culture Applications,* Maramorosch, K. and Hirumi, H., Eds., Academic Press, New York, 1979, 331.
8. **Lowry, O. H., Rosebrough, N. J., Farr, A. L., and Randall, R. J.,** Protein measurement with the Folin phenol reagent, *J. Biol. Chem.,* 193, 265, 1951.
9. **Kurtti, T. J., Munderloh, U. G., and Samish, M.,** Effect of medium supplements on tick cells in culture, *J. Parasitol.,* 68, 930, 1982.
10. **Munderloh, U. G. and Kurtti, T. J.,** Malarial parasites complete sporogony in axenic mosquitoes, *Experientia,* 41, 1205, 1985.
11. **Kurtti, T. J. and Munderloh, U. G.,** Tick cell culture: characteristics, growth requirements, and applications to parasitology, in *Invertebrate Cell Culture Applications,* Maramorosch, K. and Mitsuhashi, J., Eds., Academic Press, New York, 1982, 195.
12. **Mitsuhashi, J.,** Determination of essential amino acids for insect cell lines, in *Invertebrate Cell Culture Applications,* Maramorosch, K. and Mitsuhashi, J., Eds., Academic Press, New York, 1982, 9.
13. **Wilkie, G. E. I., Stockdale, H., and Pirt, S. V.,** Chemically-defined media for production of insect cells and viruses *in vitro, Dev. Biol. Stand.,* 46, 29, 1980.
14. **Malinoski, F. and Stollar, V.,** Inhibitors of IMP dehydrogenase prevent Sindbis virus replication and reduce GTP levels in *Aedes albopictus* cells, *Virology,* 110, 281, 1981.
15. **Wyss, C.,** Purine and pyrimidine salvage in a clonal *Drosophila* cell line, *J. Insect Physiol.,* 23, 739, 1977.
16. **Mitsuhashi, J.,** A new continuous cell line from larvae of the mosquito *Aedes albopictus* (Diptera, Culicidae), *Biomed. Res.,* 2, 599, 1981.
17. **Stollar, V. and Malinoski, F.,** The effects of adenosine and guanosine on the replication of Sindbis and Vesicular Stomatitis viruses in *Aedes albopictus* cells, *Virology,* 115, 57, 1981.
18. **Fallon, A. M. and Stollar, V.,** The biochemistry and genetics of mosquito cells in culture, *Adv. Cell Cult.,* 5, 97, 1987.

Chapter 3

THE SERUM-FREE CULTURE OF INSECT CELLS *IN VITRO*

J. Mitsuhashi and R. H. Goodwin

TABLE OF CONTENTS

I. INTRODUCTION

The importance and the necessity of developing serum-free culture media for insect cells are evident. Serum-free media are indispensable, not only for biochemical and virological studies, but also for large-scale cultures. Serum-free media may be divided into two categories: one is chemically defined media, and the other is semidefined media, or those which contain complex natural substances other than sera. The former, especially protein- (or peptide-) and lipid-free media, is extremely useful for biochemical experiments. The latter becomes important when cost, preparation, and simplicity of media are the primary considerations.

II. HISTORY

Landureau and Jolles[1] cultured a cell line from *Periplaneta americana* (EPa) in serum-free media and studied the utilization of amino acids by the cells.

Hsu et al.[2] reported that a cell line derived from *Culex tritaeniorhynchus summorosus* could be cultured in the modified 721 medium free of either insect or vertebrate sera for more than ten passages.

Kitamura et al.[3] adapted cell lines of *Culex pipiens molestus* and *Aedes aegypti* to Kitamura's BCM medium[4] containing 1% bovine plasma albumin (BPA) and two parts of medium 199 instead of calf serum. They reported that both of the cell lines grew well continuously for over 6 months in their serum-free medium.

Goodwin[5] cultured *Lymantria dispar* cells in an amino acid-based medium supplemented only with peptones instead of serum for several passages. As peptones, he used peptic peptone (NBC), liver digest (Oxoid), lactalbumin hydrolysate (LH) (Difco), and yeastolate (Difco).

Vail et al.[6] adapted *Trichoplusia ni* cells to the TNM-FH medium for which whole egg ultrafiltrate, crystalline bovine albumin, and fetal bovine serum (FBS) were deleted. They passaged the cells at least 44 times in their serum-free medium.

Hink et al.[7] cultured *T. ni* cells (TN-368) in serum-free media. In their system, Grace's medium containing 0.3% yeastolate, 0.3% LH, and 8% FBS was used as the basal medium. As a substitute for FBS, bacto-tryptose, bacto-tryptone, and bacto-peptone were evaluated singly. Bacto-tryptone was found to be the most beneficial. After fortifying the basal medium with bacto-tryptone and several serum proteins, the cells were subcultured 185 times. Furthermore, the cells could be subcultured at least 32 times in the same medium containing neither serum nor serum protein.

Pant et al.[8] obtained a cell line from the potato tuber moth (*Gnorimoschema operculella*) through the use of Mitsuhashi and Maramorosch (MM) medium.[9] They were able to adapt the cells to the MM medium containing 0.8% bovine serum albumin (BSA) in lieu of FBS and to subculture them 12 or more times at weekly intervals. In the serum-free medium, no decrease in growth rate was observed.

Goodwin and Adams[10] adapted cell lines from *Heliothis zea* and *L. dispar* to a serum-free medium. It was necessary to incorporate glycerol into the medium for prolongation of survival and subculturing. The adhesiveness and the growth of the cells were found to be affected by the concentrations of folic acid and glutamine. Later, the researchers were able to subculture the cells of *L. dispar* 80 times in another serum-free medium.[11] That medium was the modified IPL-52B-76 combination medium whose lipid mixture was replaced with a sonicated peptoliposome fraction (cholesterol, L-α-phosphatidylcholine, DL-α-tocopherol acetate). They also tried to improve their basal serum-free medium formulation. The most vigorous cell growth was obtained by the use of the IPL-52B-76 combination medium incorporating the earlier L_3 lipid mixture (5 mg methyl oleate, 25 mg Tween® 80, 4.5 mg cholesterol, and 1.75 mg α-tocopherol acetate per liter) while increasing the basal formulation to 3.6 mg/l folic acid, 2 g/l fresh glutamine, 1 g/l α-glycerophosphate, and 2 g/l glycerol.[12]

Wilkie et al.[13] developed a chemically defined medium that made it possible to continuously culture the cell lines of *Spodoptera frugiperda, A. aegypti,* and *Anopheles gambiae.*

Brooks et al.[14,15] adapted cell lines from *Blattella germanica* (UM-BGE-1 and UM-BGE-2) in modified Landureau and Jolles' S-19 medium[11] containing cholesterol instead of FBS. In this case a cholesterol emulsion made from a boiling ethanolic solution was supplied to the culture medium with an emulsion of vegetable lecithin or synthetic L-α-lecithin. The serum-free cultures were maintained for over 2 years. Interestingly, these two cell lines which adapted to the serum-free medium were highly malignant cell lines, while two other cell lines demonstrating less malignancy (by insect intrahemocoelic injection of cells) failed to adapt to the serum-free medium.[16]

Echalier[17] grew *Drosophila melanogaster* cells in D-22 medium without serum or any other protein supplementation.

Mitsuhashi[18] adapted four lepidopteran cell lines (from *Papilio xuthus, Mamestra brassicae,* and *Leucania separata*), and two dipteran cell lines (from *Aedes albopictus* and *Sarcophaga peregrina*) to serum-free MM medium. These cell lines (except the *P. xuthus* cells) were also adapted to a protein- and lipid-free medium, MTCM-1103. The growth of these cells in serum-free media was slower than in the MM medium containing 3% FBS. They could be continuously cultured in the serum-free media for an indefinite period.

Becker and Landureau[19] cultured *P. americana* cells in serum-free S-20 medium. However, they reported that a progressive decrease of cell growth occurred in that medium.

Kuno[20] cultured the cells of a mosquito, *Toxorhynchites amboinensis,* in a 1:1 mixture of L-15 and tryptose phosphate broth (TPB).

Roder[21] cultured *S. frugiperda* cells (IPLB-Sf), *Spodoptera littoralis* cells (HPB-Sl), and *M. brassicae* cells (HPB-Mb) in Gardiner and Stockdale's BML-TC/10 medium;[22] by decreasing the concentration of FBS and adding egg yolk emulsion, he was able to adapt the cells to the same medium containing 1% egg yolk instead of FBS. The cells were subcultured in this serum-free medium more than 200 times.

Goodwin[23] cultured several lepidopteran cell lines from *L. dispar* (IPLB-652A), *H. zea* (IPLB-HZ-1075), *Plodia interpunctella* (IAL-PID-2), *Malacosoma disstria* (IPRI-MD-66), and *Estigmene acrea* (BTL-EA-1179) in lipid-supplemented, serum-free RIL-2 medium. The *Choristoneura fumiferana* cell line (IPRI-CF-1) and the *Manduca sexta* cell line (MRRL-CH-34) did not adapt to the serum-free RIL-2 medium. A dipteran cell line (WR69-DM-2) from *D. melanogaster,* a hymenopteran cell line (FPMI-NI-77) from *Neodiprion lecontei,* and a coleopteran cell line (DSIR-Ha-1179) from *Heteronychus arator* were also adapted to the lipid-supplemented RIL-2 medium.

III. CHEMICALLY DEFINED MEDIA

Only a few chemically defined media have been developed for insect cells. The medium formulated by Becker and Landureau[19] and that formulated by Wilkie et al.[13] fall into this category. However, the use of these media is limited to several cell lines. Becker and Landureau's S-20 medium can support only cockroach, *P. americana,* cells (Table 1). However, if the medium was modified by changing the sugars, inorganic salts, amino acids, pH, and osmotic pressure, the modified medium (L-21) could support the growth of a lepidopteran cell line from *M. disstria.*[24] The CDM medium of Wilkie et al.[13] can be used for cultures of the cell lines from *S. frugiperda, A. aegypti,* and *A. stephensi* (Table 2). In addition to these media, M-14 medium for the primary cultures of cockroach embryonic cells.[25] and R-14 medium for *D. melanogaster* imaginal disk cultures[26] have been reported. However, these two media were not used for culturing continuous cell lines. The composition

Table 1
COMPOSITION OF LANDUREAU'S MEDIA (mg/100 ml)

Ingredients	S-19[1]	S-19B[35]	S-20[36]	L-21[24]
NaCl	850	850	847.4	—
NaHCO$_3$	36	36	—	—
KCl	105	105	104.4	300
MgCl$_2$·6H$_2$O	—	—	—	112
MgSO$_4$·7H$_2$O	126	126	123.2	412
CaCl$_2$	49	49	22.2	50
MnSO$_4$·H$_2$O	6.5	6.5	4.3	3.5
PO$_3$H$_3$	90	90	90.2	90
Glucose	300	300	400	400
Sucrose	—	—	—	500
L-α-Alanine	12	—	—	—
L-Arginine·HCl	80	80	80.1	75
L-Aspartic acid	25	100	50.6	50
L-Cysteine·HCl	26	26	72.0	37.5
L-Glutamic acid	150	150	150.1	150
L-Glutamine	30	—	—	50
Glycine	75	75	75.0	75
L-Histidine	30	30	20.2	20
L-Isoleucine	12	12	13.1	10
L-Leucine	25	25	19.7	20
L-Lysine·HCl	16	16	18.3	20
L-Methionine	50	50	25.4	25
L-Phenylalanine	20	20	9.9	10
L-Proline	75	75	63.3	65
L-Serine	8	8	3.2	40
L-Threonine	20	20	10.0	10
L-Tryptophan	20	20	10.2	10
L-Tyrosine	18	18	9.1	10
L-Valine	15	15	9.4	10
Folic acid	0.001	0.001	0.0022	0.005
d-Biotin	0.001	0.001	0.001	0.005
Choline chloride	0.04	0.04	18.15	25.0
Inositol	0.005	0.005	0.0054	0.05
Nicotinamide	0.03	0.03	0.0037	—
Ca-pantothenate	0.01	0.01	0.048	0.1
Pyridoxine·HCl	0.003	0.003	0.004	0.025
Riboflavin	0.005	0.005	0.0188	0.05
Thiamine·HCl	0.001	0.001	0.067	0.2
Cyanocobalamin	—	0.0005	0.0041	0.02
Plasma protein Fr.V	400	—	—	—
α$_2$-Macroglobulin	5.0	—	—	—
Penicillin G	5.0	5.0		12.5
Streptomycin	7.0	7.0		5.0
pH	7.4	7.4		6.5
Osmotic pressure (mOsm/kg)	440	440		320

of these media is very complicated. It would be desirable to simplify such media and modify them so as to expand their usage.

The main difficulty in formulating chemically defined media for insect cells is the fact that no specific growth factors have been isolated and identified for insect cells. A mammalian hormone, insulin, has been reported to stimulate the growth of cells from *D. melanogaster*. However, this hormone has been known to stimulate only the growth of *Drosophila* cells.

Table 2
COMPOSITION OF CDM MEDIUM

Ingredients	mg/100 ml	Ingredients	mg/100 ml
$NaH_2PO_4 \cdot 2H_2O$	114	Putrescine	0.1
$NaHCO_3$	35	Spermidine	0.1
KCl	287	Spermine-4HCl	0.1
$MgCl_2 \cdot 6H_2O$	228	Stearic acid	0.01
$MgSO_4 \cdot 7H_2O$	278	Myristic acid	0.01
$CaCl_2 \cdot 2H_2O$	132	Oleic acid	0.01
$FeSO_4(NH_4)_2SO_4 \cdot 6H_2O$	0.5	Linoleic acid	0.01
$ZnSO_4 \cdot 7H_2O$	0.044	Linolenic acid	0.01
$CuSO_4 \cdot 5H_2O$	0.039	Palmitic acid	0.01
$MnCl_2 \cdot 4H_2O$	0.035	Palmitoleic acid	0.01
α-D-Glucose	400	Arachidonic acid	0.002
L-α-Alanine	22.5	Trilinolenin	0.01
L-Arginine	55.0	Triliolein	0.01
L-Asparagine	35.0	Phosphatidylcholine	0.02
L-Asparate (K^+ salt)	45.0	Cholesterol	0.1
L-Cystine	7.5	β-Sitosterol	0.1
L-Glutamate (K^+ salt)·H_2O	82.9	Stigmasterol	0.1
L-Glutamine	60.0	Tween® 80	2.0
Glycine	65.0	Ethyl alcohol	0.2 (ml)
L-Histidine-HCl·H_2O	338.0	Riboflavin	0.02
L-Isoleucine	5.0	*p*-Aminobenzoic acid	0.2
L-Leucine	7.5	Folic acid	0.1
L-Lysine-HCl	62.5	*d*-Biotin	0.005
L-Methionine	5.0	Ca-D-pantothenate	0.12
L-Phenylalanine	15.0	Isoinositol	0.2
L-Proline	35.0	Ascorbic acid	0.02
L-Serine	55.0	Cyanocobalamin	0.1
L-Taurine	0.1	Nicotinamide	0.12
L-Threonine	17.5	Thiamine-HCl	0.2
L-Tryptophan	10.0	Pyridoxine-HCl	0.1
L-Tyrosine	7.0	Choline chloride	2.0
L-Valine	10.0	α-Tocopherol acetate	0.001
α-Amino-*n*-butyric acid	0.1	Carnitine	0.1
o-Phosphorylethanolamine	0.2	Hypoxanthine	1.0
Methylcellulose (15 cps)	200		

Note: pH = 6.3. Osmolality (mOsm/kg) = 330 with KCl.

Many vertebrate cell lines have been cultured in chemically defined media containing insulin, transferrin, and selenium as growth factors. So far, *Drosophila* cells cannot be cultured in such a medium.

For vertebrate cells, specific growth factors such as epidermal growth factor, fibroblastic growth factor, nerve growth factor, and so on have been isolated and are commercially available. No information about these types of growth factors has been obtained for insect cells. If growth factors for various insect cell types are isolated, it would facilitate the formulation of chemically defined media. Growth factors for insect cells are thought to be present in insect hemolymph, FBS, and LH. However, the isolation and purification of such factors have not yet been accomplished (cf. Chapter 1).

IV. SEMIDEFINED MEDIA

Several media from this category have been reported. These media contain one or more chemically undefined natural substances. Among these substances, hydrolysates of proteins and extracts of yeast have been used widely. LH has been used frequently in lieu of a mixture of amino acids. However, it contains both peptides of various molecular weights and growth-promoting factors.[27] Yeast extract has been used mainly as a vitamin source. It can often be replaced by a mixture of water-soluble vitamins, such as those included in Grace's medium.[28]

When the chemically undefined substances of these serum-free media are replaced with chemically defined substances, the growth of cells usually stops. MTCM-1103 medium (cf. Chapter 1, Table 8) contains LH as the sole chemically undefined component. When LH was replaced by an amino acid mixture that has based on the chemical analysis of LH, cells could not grow continuously.[27,29] This seems to suggest the presence of growth factors in the chemically undefined fraction in addition to its main components. However, isolation of these growth factors has not yet been successful.[27] Alternatively, these data may indicate that linked amino acids are required by at least some insect cells.

Goodwin and Adams[12] claimed the importance of glycerol in serum-free media. They adapted *L. dispar* cells to serum-free IPL-52-73 combination medium (Table 3); however, the cells did not grow in the same medium lacking glycerol. Goodwin[23] also formulated a serum-free medium which could support the growth of cell lines from a wide range of insects. The basal medium RIL-2 (Table 3) was prepared with a (Na/K/Mg) salt ratio of 2:1:1. This ratio was then modified for cells from insect orders other than Diptera by the addition of sodium citrate, potassium citrate, and magnesium citrate according to simple average salt ratios near those described in the literature for each order. This basal medium was supplemented with lipids that were formulated into a sterol-phospholipid peptoliposome by sterile sonication in a cup-horn sonicator.

The serum-free medium of Echalier[17] for *D. melanogaster* cell lines and that of Kitamura et al.[3] for mosquito cell lines are shown in Table 4. Besides these, Davis and Shearn's medium X[30] was reported to be successful for *D. melanogaster* imaginal disk culture, but this medium was not used in culturing continuous cell lines.

Semidefined media are generally less expensive to prepare, since purified chemicals are more costly than related crude natural substances. Serum-free MM medium (MM-SF) may be the most inexpensive formulation in this group (cf. Chapter 1, Table 8). Recently, further modifications were made in MM-SF medium so as to reduce its cost further and to simplify its preparation.[31] The mixture of inorganic salts was replaced with diluted sea water, and table sugar was used in lieu of glucose. This medium was designated as MTCM-1601. It was prepared by dissolving 7.0 g LH, 5.0 g TC-yeastolate, and 8.0 g table sugar into 1 liter of diluted sea water (one part sea water to three parts distilled water). This solution has a pH value of about 6.5 and so needs no pH adjustment. Furthermore, the medium can be sterilized by autoclaving. In this medium some lepidopteran cell lines grew more slowly than in MM-SF medium. However, most cell lines tested could be cultured continuously in this medium for an unlimited period. At present, some cell lines from Lepidoptera and Diptera can be cultured in this medium. Further modifications will be necessary to widen its application to other cell types and other insect orders.

V. ADAPTATION OF CELLS TO SERUM-FREE MEDIA

Many insect cell lines have been maintained in media containing 10% FBS. In order to eliminate FBS from the medium, one can reduce the concentration of FBS gradually (for instance, 10%, 5%, 2.5%, 1.25%, etc.), and finally to 0%. In order to achieve this, one has only to add an equal amount of serum-free medium to the cell suspension when making the subculture. In this way the cell density decreases to $1/2$; or, in other words, the cells are subcultured at a split ratio of 1:2. By repeating this procedure when the cell density has recovered to its original level, one can decrease the concentration of the serum, and the

Table 3
COMPOSITION OF IPL COMBINATION MEDIA AND RIL-2 MEDIUM
(mg/100 ml)

Ingredients	50	52	52B	73	76	RIL-2
NaCl						97.5
NaH$_2$PO$_4$·H$_2$O	116	116	116	200	200	30
NaH$_2$PO$_3$						30
NaH$_2$PO$_2$·H$_2$O						20
NaHCO$_3$	35	35	35			20
KCl	350	260	260	237.5	200	100
MgCl$_2$·6H$_2$O				62.5	62.5	50
MgSO$_4$·7H$_2$O	188	188	188	125	125	300
CaCl$_2$	50	50	66[a]	20	26.5[a]	55[a]
Glucose	700	500	500	700	700	100
Fructose						200
Sucrose	250			250		
Maltose	100	100	100	100	100	100
D-(+)-Galactose						100
D-(+)-Mannose						50
Glycerol	290	380		270		*
L-Arginine·HCl	80	80	80			80
L-Asparagine	130	130	130	100	100	70
L-Aspartic acid	130	100	100			40
Cysteine·HCl·H$_2$O						10
N-Acetyl-L-cysteine						10
L-Cystine	10	10	10			10
L-Glutamic acid	150	130	130			60
L-Glutamine	100	100	100	100	100	(200)
Glycine	20	40	40			80
L-Histidine	20	20	20			(80)[b]
Hydroxy-L-proline	80	80	80			40
L-Isoleucine	75	50	50			20
L-Leucine	25	40	40			20
L-Lysine·HCl	70	70	70			150
L-Methionine	100	100	100			120
L-Ornithine·HCl						5
L-Phenylalanine	100	100	100			80
L-Proline	50	60	60			150
DL-Serine	40	60	60			40[c]
Taurine						5
L-Threonine	20	20	20			60
L-Tryptophan	10	10	10			20
L-Tyrosine	25	25	30[d]			10
N-Acetyl-L-tyrosine						10
L-Valine	50	50	50			30
Malic acid						50
α-Ketoglutaric acid						30
Succinic acid						20
Fumaric acid						10
Citric acid·H$_2$O						10
Pyruvic acid-Na salt						20
D-Glucuronic acid						10
Folic acid	0.128	0.12	0.12	0.12	0.12	0.13
Biotin	0.016		0.016			0.021
Choline chloride						2.0
Acetyl-β-methylcholine-Cl	25.0	25.0	25.0	25.0	25.0	10.0
Isoinositol	1.04	1.0	1.0	1.0	1.0	1.0
Niacin	0.016		0.016			0.041
Ca-pantothenate	0.0008		0.016			0.116

Table 3 (continued)
COMPOSITION OF IPL COMBINATION MEDIA AND RIL-2 MEDIUM
(mg/100 ml)

Ingredients	50	52	52B	73	76	RIL-2
Pyridoxine·HCl	0.04		0.04			0.055
Pyridoxal·HCl						0.01
Pyridoxamine·HCl						0.01
p-Aminobenzoic acid	0.032		0.032			0.042
Riboflavin	0.008		0.016			0.066
Thiamine·HCl	0.008		0.016			0.216
Cyanocobalamin	0.1	0.1	0.1	0.1	0.1	0.12
DL-6,8-Thioctic acid (lipoic)						0.02
β-Carotene						0.02
L-Carnitine						0.005
Uridine						5
Inosine-5'-monophosphoric acid						30
Orotic acid						30
Chondroitin sulfate						100
α-Glycerolphosphate						15
TC-Yeastolate	400	500	500	400	500	100
Polyvinylpyrrolidone K-90	250	250	(10)	250	(100)	
$ZnCl_2$	0.004	0.004	0.004	0.00174	0.004	
$ZnSO_4·7H_2O$						0.004
$MnCl_2·4H_2O$	0.002	0.002	0.002	0.00216	0.002	
$MnSO_4·H_2O$						5
$CuCl_2·2H_2O$	0.0195	0.0195	0.02	0.00318	0.02	0.02
$(NH_4)Mo_7O_{24}·4H_2O$	0.004	0.004	0.004	0.003	0.004	0.004
$CoCl_2·6H_2O$	0.005	0.005	0.005	0.00264	0.005	0.005
$FeSO_4·7H_2O$	0.05514	0.05514	0.055	0.1103	0.05514	0.055
Na_2SeO_3						0.00175
Aspartic acid (with $FeSO_4·7H_2O$)	0.03564	0.03564		0.07129	0.03564	
Peptic peptone				600	500	
Lactalbumin hydrolysate				400		
Liver digest				400	300	50
Tryptose (Oxoid)						50
Osmotic pressure (mOsm/kg)	350	270	280	400	300	*

Note: * = not fixed (used for osmolality adjustment [0.16 ml of 50% glycerol in 100 ml of medium raises osmolality 10 mOsm]; osmolality of medium with no glycerol added is 270 mOsm).

a $+2H_2O$.
b $+HCl·H_2O$.
c   L-Serine.
d $+HCl$.

cells gradually adapt to the serum-free medium, or the system gradually selects those cells that are able to survive at the decreased serum level. Usually about three passages can be made using this method, but after that, the growth rate of the cells usually decreases, and a longer incubation period will be required for recovering the original cell density. If the cells can be subcultured ten times in this manner, the medium may be replaced with completely serum-free medium. In general, insect cells take about 3 months to adapt to serum-free media.

Brooks et al.[15] employed a similar method in adapting their *B. germanica* cell lines to a serum-free medium. The cells took 6 to 8 weeks to adjust to their serum-free formulation.

Table 4
COMPOSITION OF KITAMURA'S AND ECHALIER'S MEDIA
(mg/100 ml)

Ingredients	Kitamura et al.[3]	Echalier[17] (D-22)
NaCl	390	
NaHCO$_3$	6	
NaH$_2$PO$_4$·2H$_2$O		43
KCl	30	
KH$_2$PO$_4$	6	
CaCl$_2$	6	80
MgCl$_2$·6H$_2$O		90
MgSO$_4$·7H$_2$O		336
Glucose	120	180
Succinic acid		5.5
Malic acid		60.0
Sodium acetate·3H$_2$O		2.3
Potassium glutamate·H$_2$O		498.0
Sodium glutamate·H$_2$O		798
Glycine		500
Lactalbumin hydrolysate	600	1360
TC-Yeastolate		136
Thiamine-HCl		0.002
Riboflavin		0.002
Pyridoxine		0.002
Niacin		0.002
Calcium pantothenate		0.002
Biotin		0.001
Folic acid		0.002
Inositol		0.002
p-Aminobenzoic acid		0.002
Choline chloride		0.02
Bovine plasma albumin Fr. V	1000	
TC-199 medium	40 (ml)	
Streptomycin		10.0

The TN-368 *T. ni* cells were reported to die when they were transferred directly from serum-containing medium to serum-free medium. In order to adapt the cells to serum-free medium, they were intermediately cultured in a medium containing several serum proteins instead of whole serum.[7]

Some cell lines seem to have the ability to adapt to serum-free medium quickly. Wilkie et al.[13] reported that no adaptation period was required for transferring mosquito cell lines (*A. aegypti* and *A. gambiae*) from MM medium to his chemically defined medium (CDM). According to Roder,[21] *S. frugiperda* cells (IPLB-Sf) could be cultured in serum-free BML-TC/10 medium[22] without an adaptation period.

VI. CHARACTERISTICS OF CELLS CULTURED IN SERUM-FREE MEDIA

A. Morphology

Cells adapted to serum-free media usually cannot be distinguished morphologically from the original cell population. Hsu et al.[2] reported that *C. tritaeniorhynchus summorosus* cells adapted to his serum-free medium were similar to those cultured with FBS. Wilkie et al.[13] have reported that no change in gross cell morphology was apparent when *S. frugiperda* cells were cultured in serum-free CDM medium.

B. Growth

Once a cell line has adapted to a serum-free medium, the cells usually can be propagated in the same formulation for an unlimited period. However, the growth rate of the cells usually decreases as compared with that in the serum-containing medium of the same composition. For example, *C. tritaeniorhynchus summorosus* cells grew more slowly when they were adapted to a serum-free medium.[2] Similar results were obtained when the *T. amboinensis* cell line was adapted to Kuno's serum-free L-15-Tryptosephosphate broth combination medium.[20] When *T. ni* cells (TN-368) were cultured continuously in TNM-FH medium without FBS, their growth rate was reduced to about one half that of cells in control medium.[7]

However, there are some exceptions. No change in population doubling time was observed when *S. frugiperda* cells were cultured in CDM.[13] The *B. germanica* cell lines that adapted to serum-free medium were reported to have a growth rate comparable to the same lines in serum-containing media.[15] When the cells from *G. operculella* were cultured in MM medium whose FBS was replaced with BSA, comparable growth was obtained to the FBS-supplemented MM formulation.[8] When *S. frugiperda* cells (IPLB-Sf) were cultured in serum-free, egg yolk-containing BML-TC/10 medium,[22] the cell reproduction rate was the same as that in the medium containing 10% FBS.[21]

In serum-free culture, a higher seeding density may be required to obtain satisfactory growth. This suggests that cell growth is more density-dependent in serum-free media.[29] Brooks et al.[15] have reported that a seeding density of at least 5×10^5 cells per milliliter was required to assure that serum-free cultured *B. germanica* cells were able to reach confluence.

C. Sensitivity to Low Temperatures

Insect cells adapted to serum-free media are often sensitive to low temperature. The cells from *P. xuthus*, *M. brassicae*, *L. separata*, *A. albopictus*, and *S. peregrina* can be stored at 5°C for at least 3 months if they are maintained in serum-containing media, while in serum-free media, they are unable to survive longer than 2 weeks. These cells were successfully frozen at $-100°C$ (after adding glycerol at the final concentration of 10%), but their viability was quite reduced when they were thawed.[29] Recently a medium for freeze-storage of vertebrate cells (Freeze Medium-1) has been devised.[32] This medium prevented the common decrease observed in the viability of insect cells after freezing. However, even with this medium, insect cells that had been cultured in protein- and lipid-free media such as MTCM-1103, could not be saved.[29] Goodwin[23] reported that attempts to freeze serum-free cultured cells without lipids were unsuccessful, whereas the same cell line cultured with the peptoliposomal supplementation could be successfully stored and retrieved from liquid nitrogen.

D. Sterol Requirements

It is evident that serum-free cultures are required for the determination of the nutritional requirements of insect cells. Brooks et al.[15] reported that sterols were essential for the serum-free culturing of cell lines from *B. germanica*, and cholesterol had to be supplied as coarse granules. However, sterols were not required for the serum-free growth of cell lines from *M. brassicae*, *A. albopictus*, and *S. peregrina*.[33] Cells from these species could be continuously cultured, even under sterol-free conditions, and sterols could not be detected from the resulting cells (cf. Chapter 1, Table 9).

E. Infection with Viruses

The production of nuclear polyhedrosis virus (NPV) polyhedra seems to depend upon vigorous cell growth, and, hence, upon the serum concentration of the culture medium.[34] However, infection with viruses can occur in serum-free medium, if the medium is appropriately formulated.

Vail et al.[6] reported that serum-free cultured TN-368 cells from *T. ni* produced polyhedra of *A. californica* NPV that were comparable to those produced in serum-containing medium, and that the polyhedra produced under serum-free conditions showed normal infectivity to *T. ni* larvae.

Goodwin and Adams[11] were able to passage NPV of *L. dispar* five times in *L. dispar* cells cultured in peptoliposome-supplemented serum-free medium (IPL-52B-76), and they showed that the polyhedra formed in serum-free culture retained infectivity to *L. dispar* larvae.

Testing earlier serum-free cultures of *L. dispar* and *H. zea* cells, Goodwin and Adams noted (partial) replication of the *L. dispar*-specific NPV in the *L. dispar* cells, but no replication of the *H. zea*-specific NPV in the *H. zea* cells.[10] However, viruses were only rarely incorporated into the matrix bodies (polyhedra) unless the medium was fortified with added glycerol, increased folic acid, and undegraded glutamine. Also, no viral replication was obtained in serum-free medium lacking the lipid complex supplementation (methyl oleate, α-tocopherol acetate, cholesterol, and Tween® 80).[10]

They also could culture *L. dispar* cell lines (IPLB-LD-65Y and 652) in the serum-free IPL-50-73 combination medium. However, in this medium homologous Ld-NPV did not replicate. The addition of the synthetic peptide glycyl-L-histidyl-L-lysine at 20 µg/l (not mg/l) significantly improved cell growth, but also did not allow viral replication. When the medium was supplemented with a lipid complex L_1 (containing 5 mg methyl oleate, 25 mg Tween® 80, and 1.5 mg cholesterol per liter), both the cell growth and the viral replication were improved. In the serum-free cells, virogenic stromal development and nucleocapsid formation to envelope development around single nucleocapsids and bundles of nucleocapsids were observed. However, the formed polyhedra only rarely contained virions, notwithstanding their normal appearance. The incorporation of folic acid (24 mg/l) and fresh glutamine (2 g/l) to the lipid-supplemented serum-free IPL-52-73 combination medium increased the production of polyhedra when Ld-NPV was inoculated.[12] Further improvements in viral replication (the first regular occlusion of virions within the polyhedra) were attained by modification of the peptone-based half of the combination medium (IPL-73), in which the LH was deleted and the yeastolate was increased from 4 to 5 g/l; the peptic peptone and liver digest remained unchanged (IPL-76). When this 52-76 combination medium was supplemented with 5 g/l of α-glycerophosphate in addition to the 3.7 g of glycerol, polyhedra were formed more quickly, and virion envelopes appeared to be more substantial than previously.[12]

Wilkie et al.[13] have also reported that *S. frugiperda* cells grown in CDM medium supported the replication of Ac-NPV, and the virus replicated over four passages *in vitro*.

Roder[21] also reported that *Autographa californica* NPV replicated in *S. frugiperda* cells (IPLB-Sf) cultured in a serum-free medium, and that the resultant polyhedra showed unaltered morphology and virulence.

In an early study of a mammalian virus infection, the cells from *T. amboinensis* that were adapted to a serum-free medium supported the replication of dengue virus.[20]

F. Other Characteristics

The cells of *B. germanica* that had adapted to a serum-free medium were reported to be more fragile than they were when grown in serum-containing medium.[15] These cells therefore required rather gentle handling when they were subcultured. The cells from another cockroach (*P. americana*) cell line were also easily damaged when subcultured in serum-free medium.[19]

Brooks and Tsang[14] examined the lipid uptake by *B. germanica* cells, since the cells required supplementation with cholesterol and lecithin when they were cultured in serum-free medium. Analyses of the spent medium after 1 week of culture suggested that approximately the same amount (23 to 29 µg/ml) of lipid or sterol-like substance was removed by the cells from the basal medium as from either the serum-free cholesterol- and lecithin-containing medium or the FBS-containing medium. Since the cells died within a week in the basal medium, in spite of the fact that the basal medium contained 54 µg/ml of lipids material (presumably from the yeastolate and LH), the effect of the coarsely dispersed cholesterol was thought to be protective rather than directly nutritional. If cholesterol was finely dispersed by sonication, the protective effect was lost, and the growth rate of the cells declined. The medium in which Tween® 80 was used as a dispersing agent was found to be toxic to the cells.

Some cell lines which have adapted to serum-free media seem to become sensitive to antibiotics that are routinely used for culturing them in serum-containing media. For example, *P. xuthus* cells were cultured in antibiotic-free MM-SF medium but did not survive in the same medium with an antibiotic mixture (100 µg dihydrostreptomycin sulfate, 10 U crystalline penicillin G potassium, 100 µg kanamycin sulfate, and 10 µg novobiocin per milliliter). This tendency was more evident in protein- and lipid-free medium. *A. albopictus* cells (NIAS-AeAl-2) could multiply in MM-SF medium containing the above antibiotic mixture, but the cells that adapted to the MTCM-1103 medium free of proteins and lipids ceased their growth when the antibiotics mixture was added to the medium. However, the mosquito cells were able to multiply in the MTCM-1103 medium containing only neomycin at the final concentration of 10 µg/ml.[29] It was not ascertained which antibiotics in the mixture were detrimental in the serum-free medium formulations.

Related to this, it is interesting that the *D. melanogaster* cell line K was reported to be unstable in karyotype after adaptation to serum-free D-22 medium.[17]

REFERENCES

1. **Landureau, J. C. and Jollés, P.,** Étude des exigences d'une lignée de cellules d'insectes (souche EPa). 1. Acides amines, *Exp. Cell Res.*, 54, 391, 1969.
2. **Hsu, S. H., Li, S. Y., and Cross, J. H.,** A cell line derived from ovarian tissue of *Culex tritaeniorhynchus summorosus* Dyar, *J. Med. Entomol.*, 9, 86, 1972.
3. **Kitamura, S., Imai, T., and Grace, T. D. C.,** Adaptation of two mosquito cell lines to medium free of calf serum, *J. Med. Entomol.*, 10, 488, 1973.
4. **Kitamura, S.,** Establishment of cell line from *Culex* mosquito, *Kobe J. Med. Sci.*, 16, 41, 1970.
5. **Goodwin, R. H.,** Insect cell growth on serum-free media, *In Vitro*, 12, 303, 1976.
6. **Vail, P. V., Jay, D. L., and Romine, C. L.,** Replication of the *Autographa californica* nuclear polyhedrosis virus in insect cell lines grown in modified media, *J. Insect Physiol.*, 28, 263, 1976.
7. **Hink, W. F., Strauss, E. M., and Lynn, D. E.,** Growth of TN-368 insect cells in serum-free media, *In Vitro*, 13, 177, 1977.
8. **Pant, U., Mascarenhas, A. F., and Jagannathan, V.,** In vitro cultivation of a cell line from embryonic tissue of potato tuber moth, *Gnorimoschema operculella* (Zellev), *Indian J. Exp. Biol.*, 15, 244, 1977.
9. **Mitsuhashi, J. and Maramorosch, K.,** Leafhopper tissue culture: embryonic, nymphal, and imaginal tissues from aseptic insects, *Contrib. Boyce Thompson Inst.*, 22, 435, 1964.
10. **Goodwin, R. H. and Adams, J. R.,** Serum-free media for nutritional analysis and viral replication in lepidopteran insect cells, *In Vitro*, 14, 351, 1978.
11. **Goodwin, R. H. and Adams, J. R.,** Liposome incorporation of factors permitting serial passage of insect viruses in lepidopteran cells grown in serum-free medium, *In Vitro*, 16, 222, 1980.
12. **Goodwin, R. H. and Adams, J. R.,** Nutrient factors influencing viral replication in serum-free insect cell line culture, in *Invertebrate Systems in Vitro*, Kurstak, E., Maramorosch, K., and Dübendorfer, A., Eds., Elsevier/North-Holland, Amsterdam, 1980, 493.
13. **Wilkie, G. E. I., Stockdale, H., and Pirt, S. T.,** Chemically-defined media for production of insect cells and viruses in vitro, *Dev. Biol. Stand.*, 46, 29, 1980.
14. **Brooks, M. A. and Tsang, K. R.,** Replacement of serum with cholesterol and lipids for cell lines of *Blattella germanica*, *In Vitro*, 16, 222, 1980.

15. **Brooks, M. A., Tsang, K. R., and Freeman, F. A.,** Cholesterol as a growth factor for insect cell lines, in *Invertebrate Systems in Vitro,* Kurstak, E., Maramorosch, K., and Dübendorfer, A., Eds., Elsevier/North-Holland, Amsterdam, 1980, 67.

16. **Tsang, K. R. and Brooks, M. A.,** Comparison of serum dependency of transformed insect cell lines of different degrees of malignancy, *In Vitro,* 19, 775, 1983.

17. **Echalier, G.,** In vitro established lines of *Drosophila* cells and applications in physiological genetics, in *Invertebrate Tissue Culture, Applications in Medicine, Biology and Agriculture,* Kurstak, E. and Maramorosch, K., Eds., Academic Press, New York, 1976, 131.

18. **Mitsuhashi, J.,** Continuous cultures of insect cell lines in media free of sera, *Appl. Entomol. Zool.,* 17, 575, 1982.

19. **Becker, J. and Landureau, J. C.,** Specific vitamin requirements of insect cell lines (*P. americana*) according to their tissue origin and in vitro conditions, *In Vitro,* 17, 471, 1981.

20. **Kuno, G.,** Cultivation of mosquito cell lines in serum-free media and their effects on Dengue virus replication, *In Vitro,* 19, 707, 1983.

21. **Roder, A.,** Development of a serum-free medium for cultivation of insect cells, *Naturwissenschaften,* 69, 92, 1982.

22. **Gardiner, G. R. and Stockdale, H.,** Two tissue culture media for production of lepidopteran cells and nuclear polyhedrosis virus, *J. Invertebr. Pathol.,* 25, 363, 1975.

23. **Goodwin, R. H.,** Growth of insect cells in serum-free media, in *Techniques in the Life Sciences, Cell Biology,* Vol. C 1, Elsevier County Clare, Ireland, 1985, 28.

24. **Landureau, J. C.,** Insect cell and tissue culture as a tool for developmental biology, in *Invertebrate Tissue Culture, Applications in Medicine, Biology and Agriculture,* Kurstak, E. and Maramorosch, K., Eds., Academic Press, New York, 1976, 101.

25. **Marks, E. P., Reinecke, J. P., and Coldwell, J. M.,** Cockroach tissue in vitro: a system for the study of insect cell biology, *In Vitro,* 3, 85, 1967.

26. **Robb, J.,** Maintenance of imaginal discs of *Drosophila melanogaster* in chemically defined media, *J. Cell Biol.,* 41, 876, 1969.

27. **Nagasawa, H., Mitsuhashi, J., and Suzuki, A.,** Growth factors in the lactalbumin hydrolysate to the cells of the flesh fly, *Sarcophaga peregrina* in *Invertebrate and Fish Tissue Culture,* Kuroda, Y., Kurstak, E., and Maramorosch, K., Eds., Japan Sci. Soc. Press, Springer-Verlag, Tokyo, 1987, 29.

28. **Grace, T. D. C.,** Establishment of four strains of cells from insect tissue grown in vitro, *Nature (London),* 195, 788, 1962.

29. **Mitsuhashi, J.,** unpublished data, 1987.

30. **Davis, K. T. and Shearn, A.,** In vitro growth of imaginal disks from *Drosophila melanogaster, Science,* 196, 438, 1977.

31. **Mitsuhashi, J.,** Simplification of media and utilization of sugars by insect cells in cultures, in *Invertebrate and Fish Tissue Culture,* Kuroda, Y., Kurstak, E., and Maramorosch, K., Eds., Japan Sci. Soc. Press., Springer-Verlag, Tokyo, 1987, 15.

32. **Ohno, T., Kurita, K., Abe, S., Tzimori, N., and Ikawa, Y.,** A simple freezing medium for serum-free cultured cells, *Cytotechnology,* 1, 257, 1988.

33. **Mitsuhashi, J., Nakasone, S., and Horie, Y.,** Sterol-free eukaryotic cells from continuous cell lines of insects, *Cell Biol., Int. Rep.,* 7, 1057, 1983.

34. **Watanabe, H.,** Effect of the concentration of fetal bovine serum in a culture medium on the susceptibility of *Bombyx mori* cells to a nuclear polyhedrosis virus, *Appl. Entomol. Zool.,* 22, 397, 1987.

35. **Landureau, J. C.,** Étude des exigences d'une lignée de cellules d'insectes (souche EPa). II. Vitamines hydrosolubles, *Exp. Cell Res.,* 54, 399, 1969.

36. **Landureau, J. C. and Grellet, P.,** Nouvelles techniques de culture in vitro de cellules d'insectes et leurs applications, *C. R. Acad. Sci. (Paris),* D-274, 1372, 1972.

Chapter 4

REPAIR OF SINGLE-STRAND BREAKS IN DNA FROM CULTURED LEPIDOPTERAN CELLS EXPOSED TO GAMMA RADIATION*

D. A. Stock and P. M. Achey

TABLE OF CONTENTS

*   Florida Agricultural Experiment Station Journal Series No. 4762.

I. INTRODUCTION

A wide range of resistance to radiation damage has been observed among eukaryotic cells.[1,2] As a group, insect cells have greater resistance than most cells. The D_o for mammalian cells approximates 2 Gy, while insect cells have D_{37} values ranging from 10 Gy in the Diptera to 400 Gy in the Lepidoptera.[2] Differences in radiosensitivity might result from differences in intrinsic sensitivity to initial radiation damage in DNA or be due to differences in efficiencies of DNA repair by the cells. Evidence that DNA is a critical target from cellular radiation damage includes the use of focused microbeams of ionizing radiation to show that the greatest killing efficiency results when the damage is in the chromosomes.[3,4] Additionally, DNA appears to be the critical molecular target, particularly under hypoxic conditions. This is indicated by the high killing efficiency during transmutation of radio-isotopes incorporated into the DNA compared to the killing efficiency of external irradiation, the presence of a greater fraction of nonrepairable DNA damage after neutron radiation compared to gamma radiation,[5] and the enhancement of cell killing by electron-affinic drugs that serve to enhance energy transfer to DNA molecules leading to strand breaks in the DNA "backbone".[6]

Knowledge of biochemical damage to DNA in cultured insect cells from ionizing radiation and the repair of this damage is limited and restricted primarily to *Drosophila melanogaster* of the order Diptera.[7] It has been shown that insect cells repair radiation damage but that their efficiencies vary widely.[2] Comparing gamma-irradiated cultured cells of *Ceratitis capitata* (Mediterranean fruit fly) and *Aedes albopictus* (forest day mosquito) showed that cell survival was greater in cells with a greater chromosome number.[8] Cultured cells from meiotic recombination mutants (*mei*) of *D. melanogaster* have reduced repair of single-strand and double-strand breaks in their DNA, suggesting a role for the *mei* gene product in DNA repair.[9-11]

Because of the great radioresistance of lepidopteran cells and the agricultural importance of many lepidopteran pests, we began a study of the mechanism of DNA damage from ionizing radiation and its repair in cultured imaginal wing disk cells from *Plodia interpunctella* (Indian meal moth) larvae. One of the classes of molecular damage in DNA from radiation exposure is single-strand breaks, for which a sensitive assay was developed first in experiments with bacteria[12] and later extended to eukaryotic cells.[13] Unfortunately, this assay requires separation of damaged and undamaged radioactively labeled DNA in sucrose velocity sedimentation gradients, a procedure which is difficult or not practical when working with tissue from intact organisms and when using low doses of ionizing radiation. Instead, we measured single-strand breaks by using horizontal alkaline agarose-slab gel electrophoresis and detected the DNA by means of ethidium bromide.[14] The technique has been refined by Sutherland et al.[15] and has been used to quantify pyrimidine dimers in experiments on human skin exposed to UVB radiation.[16]

II. MATERIALS AND METHODS

P. interpunctella cells grown in Grace's medium containing 10% fetal bovine serum (FBS) as described previously[17] were supplied by Dr. H. Oberlander, Insect Attractants, Behavior, and Basic Biology Research Laboratory, U.S. Department of Agriculture, Agricultural Research Service, Gainesville, FL. Cells were seeded at 250,000 cells per milliliter in 25-ml plastic tissue culture flasks, then incubated 1 week at 26°C. After incubation the cells had formed a confluent monolayer containing 3 to 5 million cells per milliliter in a 5.5-ml volume.

Cultured cells were removed from the wall of the flask by agitating the medium with a Pasteur pipette. The suspension of cells was transfered to a 13 × 100-mm test tube.

Irradiation was from a ^{60}Co custom-designed gamma irradiator[18] with a dose rate of 3636 rad/min. The cell suspensions were bubbled with dioxygen continuously during irradiation at a flow rate of 60 cc/min. The total volume of the cell suspension was 5 ml at the beginning of the irradiation exposure. At appropriate times of exposure, 0.25-ml samples were removed and transferred to tubes held on ice. Some of the samples were incubated 60 min in a water bath at 24°C to determine if repair of DNA occurred in the irradiated cells. The remaining samples were kept on ice. The cell suspensions were centrifuged; then the cell pellets were resuspended in 40 μl A-1 buffer.[19] A-1 buffer contains 50 mM Tris, 10 mM EDTA, and 400 mM NaCl and has a pH of 8. In order to lyse the cells, 1μl 5% sodium dodecyl sulfate (SDS) was added, followed immediately by the addition of either 1.6 μl of a 1-mg/ml protease IV (Sigma Chemical Co., St. Louis) solution which had been predigested at 37°C for 60 min, or 10 μl of a 25%-glycerol, 0.5 N-NaOH, and 0.125%-bromocresol green solution. Samples treated with protease were incubated 60 min at 37°C. Protease was used to digest nucleases and nucleoprotein associated with DNA to permit better electrophoretic separation of the DNA strands. When protease was not used, the alkaline dye solution served to inhibit nucleases. After incubation with protease, 10 μl of the alkaline dye solution was added. A 20-μl sample of the resulting solution was placed in the well of a horizontal agarose gel containing 0.3% agarose, 0.3 N NaOH, and 0.01 M trisodium EDTA. Electrophoresis was carried out for 20 h at room temperature using a voltage gradient of 0.75 V/cm. Electrophoresis buffer contained 0.3 N NaOH and 0.01 M trisodium EDTA. After electrophoresis the gel was carefully removed (dilute gels are subject to tearing) and placed in a tray. The gel was covered with 500 ml 0.1 M Tris, pH 8, and allowed to neutralize for 30 min. The buffer was removed by aspiration, and the gel was covered with 500 ml 0.1 M Tris, pH 8, containing 0.5 μg/ml ethidium bromide. After incubation for 30 min to allow staining of the DNA, the gel was placed on an ultraviolet (UV) light table (300 nm transillumination) and photographed using Type 55 P/N Polaroid® film.

III. RESULTS

As the radiation dose was increased, the number of single-strand breaks in the DNA of cultured *P. interpunctella* cells increased. This is evident by the increased streaking seen in lanes containing DNA from cells receiving larger doses of radiation (Figure 1). Also observable in Figure 1 is the repair (rejoining) of single-strand breaks in cells incubated 60 min at 24°C before DNA extraction. This is particularly noticeable by the reduced streaking of these samples and the accumulation of material nearer to the origins of the lanes containing DNA from incubated cells receiving radiation doses of 50, 200, and 400 Gy.

Alternatively, we scanned the negative of a photograph of the gel using a gel scanner attachment and chart recorder attached to a Model 260 Gilford® spectrophotometer, set at a wavelength of 550 nm. Tracings of 0-, 200-, and 400-Gy irradiated samples with and without postirradiation incubation to allow repair are plotted in Figure 2. The direction of migration of the DNA samples is from left to right. Both the 200- and 400-Gy samples which had been incubated after irradiation migrated more slowly than the unincubated samples, indicating that repair of single-strand breaks occurred.

IV. DISCUSSION

The results of this study show that repair of single-strand breaks in DNA occurs in cultured insect cells. Our results using *P. interpunctella* cells correspond with the observation of repair in wild-type eukaryotic cells from the yeast *Saccharomyces cerevisiae*[20] and *D. melanogaster*.[21] Repair of single-strand breaks also has been reported in carrot protoplasts,[22] mammalian cells,[13] and various prokaryotic cells.[12]

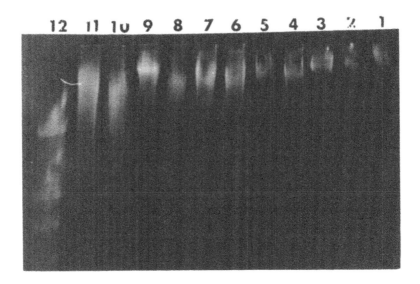

FIGURE 1.   Horizontal alkaline agarose gel electrophoresis of DNA from cultured *P. interpunctella* cells. From right to left, the lanes contain DNA from (1) nonirradiated cells; (2) 2 krad; (3) 2 krad, incubated; (4) 5 krad; (5) 5 krad, incubated; (6) 10 krad; (7) 10 krad, incubated; (8) 20 krad; (9) 20 krad, incubated; (10) 40 krad; (11) 40 krad, incubated; (12) molecular weight reference DNA consisting of lambda-DNA digested with *Hind*III endonuclease. Fragment single-strand molecular weights are (in megadaltons) 7.75, 3.20, 2.18, 1.42, 0.726.

Cavalloro et al.[8] restricted their studies to damage of nuclear structures in cultured insect cells from radiation exposure. By comparing the radiosensitivity of *A. albopictus* cells with *C. capitata* cells, which have twice the number of chromosomes as the cells of *A. albopictus*, they found *A. albopictus* cells more radiosensitive than *C. capitata* cells. Through use of cytophotometric measurements, the DNA content of *C. capitata* cells was observed to become irregular after gamma radiation, and the degree of irregularity increased with increasing radiation.[8] Indirectly, our results corroborate their study in that we found that as the radiation dose to *P. interpunctella* cells was increased, single-strand breaks were more numerous in the DNA.

The discovery of mutants of *D. melanogaster* which have altered meiotic recombination (*mei*) or which are sensitive to methyl methanesulfonate (*mus*) has provided opportunities for examining mechanisms for DNA repair.[7,21] Radiosensitive mutant strains of *D. melanogaster* were able to repair single-strand breaks as readily as wild-type cells.[21] *D. melanogaster* cells could repair double-strand breaks in DNA; however, two of the mutant strains in this study had reduced ability to repair double-strand breaks.[11] Spermatogonia of one *D. melanogaster* mutant (*mei*-9) exhibited lower mutation frequency in males following irradiation under anoxia and posttreatment with dioxygen compared to spermatogonia posttreated with dinitrogen.[23] Hence, the repair defect in *mei*-9 does not block repair of dioxygen-dependent postirradiation repair. Postirradiation atmospheres had no effect on repair processes in spermatogonia of two other mutants, *mei*-41 and *mus*-101.[23] *Ebony* mutants of *D. melanogaster* are sensitive to both UV and X-irradiation and have defects in repair of single-strand breaks in DNA and photoreactivation repair.[24] Males (*mei*-9[a] and *mei*-9[1.1]) deficient in excision repair formed recessive lethal mutations after X-irradiation similar to repair proficient males, whereas males (*w mus*[1]101[d]) deficient in postreplication repair had a lower formation of recessive lethals than repair proficient males, particularly at low dose rates.[25] Our understanding of repair processes is in its infancy for eukaryotic organisms, even in a well-studied species such as *D. melanogaster*.

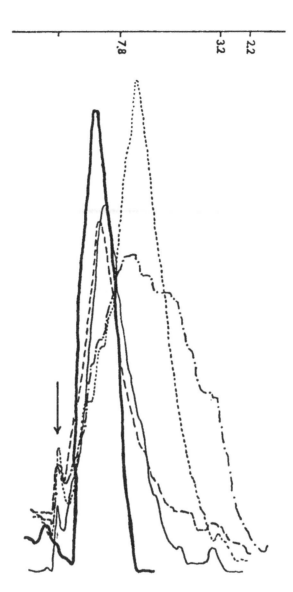

FIGURE 2. Absorption profiles for selected lanes of negative from photograph of gel in Figure 1. The lanes scanned are control (————); 20 krad (· · · ·); 20 krad, incubated (— — — — —); 40 krad (— · — · —); 40 krad, incubated (————). The locations of single-strand molecular weight standards are indicated in megadaltons. The well edges are marked by the down-pointing arrow.

Our use of alkaline agarose gels and omission of a phenol-extraction step has permitted us to detect DNA repair using lower radiation exposure (20 Gy as a lower limit) than has been typically used for studying single-strand breaks, because the DNA is subjected to less shearing during preparation. Experiments using alkaline sucrose or neutral sucrose density gradients has typically required exposure of 100 to 1000 Gy.[20-22] This is because randomly induced DNA breaks resulting from the preparation and analysis procedures are superimposed on the radiation-induced strand breaks.

Although it is generally conceded that DNA is the most sensitive target for radiation damage in the cell,[26] irradiation of intact cells could lead to damage of other cellular structures which could affect cellular activity. Therefore, we looked for a method which would permit us to irradiate DNA separately, then introduce it into nonirradiated cells and observe if any repair of the introduced DNA occurred. The development of recombinant DNA and cell transfection techniques have permitted us to observe damage of DNA independent of a cell. We are now in the process of applying these techniques to study mechanisms of repair of radiation-damaged DNA in a variety of insect cells.

Our interest in radiation effects on insect cells results from a desire to apply radiation to the control of economically damaging insect populations in a manner similar to the biological control of screw worm and Mediterranean fruit flies, using radiation-sterilized males. Different levels of repair of single-strand breaks occur in male mouse germ cells, depending on the maturity level: 50% efficiency of break rejoining in spermatogonia and essentially 0% efficiency in spermatozoa.[27] This important difference in repair efficiencies suggests that the success of a controlled, sterile-male release program could be enhanced if the somatic cells, such as those studied here, possessed the ability to repair strand breaks whereas the spermatozoa did not, and the radiation conditions were chosen to favor repair of radiation damage to DNA in the somatic cells. This could be achieved by choosing an appropriate chemical environment during radiation exposure, possibly by using an oxic or anoxic environment, so that the chemical nature of the DNA damage allowed it to be more effectively repaired. It has been observed that the fraction of repairable DNA strand breaks is larger for irradiation under anoxic conditions than for irradiation in dioxygen;[2,28] however, for certain *D. melanogaster* cells the opposite seems to occur.[23] Choosing the appropriate radiation conditions with consideration given to the repair of the damage could result in more biologically active males which at the same time compete with the males in the wild.

REFERENCES

1. **Casarett, A. P.,** *Radiation Biology,* Prentice-Hall, Englewood Cliffs, NJ, 1968.
2. **Koval, T. M.,** Intrinsic resistance to the lethal effect of X-irradiation in insect and arachidonic cells, *Proc. Natl. Acad. Sci. U.S.A.,* 80, 4752, 1983.
3. **Ord, T. and Danielli, J. F.,** The site of damage in amoebae exposed to X-rays, *Q. J. Microsc. Sci.,* 97, 29, 1956.
4. **Zirkle, R. E.,** Partial-cell irradiation, *Adv. Biol. Med. Phys.,* 5, 103, 1957.
5. **Ahnstrom, G. and Ehrenberg, L.,** The nature of the target in the biological action of ionizing radiations, *Adv. Biol. Med. Phys.,* 17, 129, 1980.
6. **Greenstock, C. L. and Whitehouse, R. P.,** Radiosensitizers as probes of DNA damage and cell killing, *Int. J. Radiat. Biol.,* 48, 701, 1985.
7. **Boyd, J. B., Snyder, R. D., Harris, P. V., Presley, J. M., Boyd, S. F., and Smith, P. D.,** Identification of a second locus in *Drosophilia melanogaster* required for excision repair, *Genetics,* 100, 239, 1982.
8. **Cavalloro, R., Rozi, G., and Zaffaroni, G.,** Radiosensitivity and DNA content of an in vitro cell line of the Mediterranean fruit-fly, in *Invertebrate Systems In Vitro,* Kurstak, E., Maramorosch, K., and Dübendorfer, A., Eds., Elsevier/North-Holland, Amsterdam, 1980, 79.
9. **Boyd, J. B. and Harris, P. V.,** Mutants partially defective in excision repair at five autosomal loci in *Drosophila melanogaster, Chromosoma,* 82, 249, 1981.
10. **Boyd, J. B. and Presley, J. M.,** Repair replication and photorepair of DNA in larvae of *Drosophila melanogaster, Genetics,* 77, 687, 1974.
11. **Dezzani, W., Harris, P. V., and Boyd, J. B.,** Repair of double-strand DNA breaks in Drosophila, *Mutat. Res.,* 92, 151, 1982.
12. **McGrath, R. A. and Williams, R. W.,** Reconstruction *in vivo* of irradiated *Escherichia coli* deoxyribonucleic acid; the rejoining of broken pieces, *Nature (London),* 212, 534, 1966.
13. **Lett, J. T., Caldwell, I., Dean, C. J., and Alexander, P.,** Rejoining of X-ray induced breaks in the DNA of leukemia cells, *Nature (London),* 214, 790, 1967.
14. **Achey, P. M., Woodhead, A. D., and Grist, E.,** Measurement of pyrimidine dimers in unlabeled DNA, in *DNA Repair: A Laboratory Manual of Research Procedures,* Vol. 2, Friedberg, E. C. and Hanawalt, P. C., Eds., Marcel Dekker, New York, 1983, 109.

15. **Sutherland, J. C., Montelone, D. C., Trunk, J. C., and Ciarrocchi, G.,** Two-dimensional, computer controlled film scanner: quantitation of fluoroscence from ethidium bromide-stained DNA gels, *Anal. Biochem.,* 139, 390, 1984.
16. **Freeman, S. E., Gange, R. W., Matzinger, E. A., and Sutherland, D. M.,** Higher pyrimidine dimer yields in skin of normal humans with higher UVB sensitivty, *J. Invest. Dermatol.,* 86, 34, 1986.
17. **Lynn, D. E. and Oberlander, H.,** Development of cell lines from imaginal wing discs of Lepidoptera, *In Vitro,* 17, 208, 1981.
18. **Hanrahan, R. J.,** A ^{60}Co gamma irradiator for chemical research, *J. Appl. Radiat. Isot.,* 13, 234, 1962.
19. **Achey, P. M., Woodhead, A. D., and Setlow, R. B.,** Photoreactivation of pyrimidine dimers in DNA from thyroid cells of the teleost, *Paecilia formosa, Photochem. Photobiol.,* 29, 305, 1979.
20. **Mowat, M. R. A., Jachymczyk, W. J., Hastings, P. J., and von Barstel, R. C.,** Repair of gamma-ray induced DNA strand breaks in the radiation-sensitive mutant *rad* 18-2 of *Sacchromyces cerevisiae, Mol. Gen. Genet.,* 189, 256, 1983.
21. **Boyd, J. B. and Setlow, R. B.,** Characterization of postreplication repair in mutagen-sensitive strains of *Drosophila melanogaster, Genetics,* 84, 507, 1976.
22. **Howland, G. P., Hart, R. W., and Yette, M. L.,** Repair of DNA strand breaks after gamma-irradiation of protoplasts isolated from cultured wild carrot cells, *Mutat. Res.,* 27, 81, 1975.
23. **Eeken, J. C. J. and Sobels, F. H.,** Studies on mutagen-sensitive repair deficiencies in spermatids, spermatocytes, and spermatogonia irradiated in N_2 or O_2, *Mutat. Res.,* 149, 409, 1985.
24. **Ferro, W.,** Studies on mutagen-sensitive strains of *Drosophila melanogaster.* V. Biochemical characterization of a strain (ebony) that is UV and X-ray sensitive and deficient in photorepair, *Mutat. Res.,* 149, 399, 1985.
25. **Sankaranarayanan, K. and Ferro, W.,** Studies on mutagen-sensitive strains of *Drosophila melanogaster.* VII. Effects of repair deficiency in males on X-ray-induced sex-linked recessive lethals in spermatozoa, *Mutat. Res.,* 149, 415, 1985.
26. **Alper, T.,** *Cellular Radiobiology,* Cambridge University Press, Cambridge, 1979.
27. **Ono, J. and Okada, S.,** Radiation-induced DNA single-strand scission and its rejoining in spermatogonia and spermatozoa of mouse, *Mutat. Res.,* 43, 25, 1947.
28. **Sapora, O., Fielden, E. M., and Loverock, P. S.,** The application of rapid lysis techniques in radiobiology. I. The effect of oxygen and radiosensitizers on DNA strand break production and repair in *E. coli* B/r, *Radiat. Res.,* 64, 431, 1975.

Chapter 5

INCORPORATION AND METABOLISM OF *N*-ACETYLGLUCOSAMINE BY A LEPIDOPTERAN CELL LINE

Edwin P. Marks, Eric B. Jang, and Randel Stolee

TABLE OF CONTENTS

I. INTRODUCTION

The production of cuticle and biosynthesis of chitin by cultured insect tissue has been demonstrated in several systems.[1] Biosynthesis of chitin in cell-free systems derived from insect tissues has also been accomplished.[2,3] Dübendorfer et al.[4] reported deposition of a cuticle (presumably containing chitin) by vesicles formed in freshly dissociated primary cell cultures of *Drosophila*. We have recently found evidence that cuticle deposition can occur in three established insect cell lines.[5] This occurred consequent to freezing and long-term storage. These findings prompted us to investigate incorporation and metabolism of the chitin precursor *N*-acetylglucosamine by the MRRL-CH cell line derived from embryos of the lepidopteran insect, *Manduca sexta* L.

II. METHODS

A. The Cell Lines

Cell lines MRRL-CH and MRRL-CH2 were derived from embryos of *M. sexta*[6] grown in Grace's modified medium[7] supplemented with 15% fetal bovine serum (FBS). Cells used were frozen in the 34th passage in 1973 and stored in liquid nitrogen until 1982. When they were removed from storage and cultured, spheres composed of cuticle-like material were found among the cells. The recovered cells were grown in Grace's medium, subcultured at 7-d intervals, and maintained in a dark incubator at 27°C.

The UMBGE-2 cell line derived from cockroach embryos was obtained from Kurtti of the University of Minnesota[8] and adapted to Grace's modified medium supplemented with 10% FBS. The adapted line was maintained at 27°C and subcultured at weekly intervals. In 1977, after 22 passages, the adapted cells were frozen and stored in liquid nitrogen. They were recovered in 1978, and cuticle spheres were found in the developing cultures.

B. Cuticle Spheres

A 0.5-ml aliquot of cells containing cuticle spheres was added to 5 ml of hyamine hydroxide and incubated for 24 h. The digest was centrifuged and washed three times, and the pellet was fixed in phosphate-buffered gluteraldehyde (3%, pH 7.2). It was then postfixed in 2% osmium tetraoxide, dehydrated in a graded series of acetone, embedded in epon, sectioned, stained with lead citrate and uranyl acetate, and examined by electron microscopy.

C. Uptake of *N*-Acetylglucosamine

Uptake of *N*-acetylglucosamine (GlcNAc) was measured in two experiments. For both, 24 flasks of 25 cm^2 containing 5 ml of medium were seeded with 150,000 cells. They were incubated for 2 weeks and then dosed with 5 μl (0.5 μCi) of ^{14}C-labeled GlcNAc in ethanol/water (9:7) (57.2 mCi/mM) for 24 h. In the first experiment, sets of three flasks were randomly selected at intervals of 1, 3, 6, 10, 24, 28, 32, and 48 h, harvested as described above, and counted for radioactivity. In the second experiment, sets of three flasks were taken at 24, 48, 58, 72, 96, 120, 144, and 168 h.

Ability of the cells to retain GlcNAc taken up in two flasks was measured by washing the monolayer of dosed cells three times with Puck saline A solution containing 4×10^3 M cold GlcNAc and refilled with fresh medium,. Samples of the medium were counted hourly for the first 7 h and again at 22 h. The monolayer was washed, removed by scraping, and counted by liquid scintillation.

D. Inhibition of Incorporation of GlcNAc

Plastic flasks (150 cm^2) containing 30 ml of Grace's modified medium containing 15% FBS were seeded with 9×10^5 cells from the MRRL-CH line. Experiments were run on the 14th or 15th day after seeding. The cells were harvested by scraping, then centrifuged, and resuspended in 8 ml of the supernate. A 0.5-ml aliquot was removed for scintillation counting. One milliliter of the vortexed cell suspension was placed in each of six centrifuge tubes, and 4.0 ml of the supernatant medium was added to each tube.

Tubes 1 through 3 received the competing substrate or inhibitor dissolved in 5-μl dimethyl sulfoxide (DMSO). Final concentrations of the experimental compound were oligomycin, antimycin A, 10^{-5} M; GlcNAc, N-acetyl-D-galactosamine, and N-acetyl-D-mannosamine, 5 × 10^{-4} M. Five microliters of DMSO were placed in tubes 4 through 6 as the controls.

All six tubes were vortexed briefly and incubated at room temperature for 1 h. Each tube was dosed with 5 μl (0.5 μCi) ^{14}C-GlcNAc in EtOH/H$_2$O (35.6 mCi/mM). The tubes were vortexed briefly and incubated at room temperature for 3 h, then spun at 2500 rpm for 5 min on a tabletop centrifuge. The cells were centrifuged and washed twice with 3-ml portions of Puck saline solution A. The cells were transferred to scintillation vials containing 10 ml of Beckman® HP scintillation cocktail and counted on a liquid counter.

E. Low Temperature Experiments

Low temperature experiments varied slightly from the above procedure. No DMSO was added to the tubes, and the tubes were either left at room temperature or placed in an ice bath (0.5 to 2.0°C) and incubated at indicated temperatures for 3 h. In another experiment, both sets of tubes were incubated at 2°C, but the dosage varied in values of 1, 3, 5, 8, 10, and 15 μl.

F. Incorporation Experiments

Extraction of cells — Cells were dosed in the flasks containing 10 μl of ^{14}C-GlcNAc and incubated for 72 h. The monolayer of treated cells was rinsed four times with Puck saline A solution and harvested by vigorous shaking. The cells were centrifuged, and the pellet was washed twice in the same solution and frozen at −30°C to disrupt the membranes. When thawed, they were extracted in either ether/water (3:1) or chloroform/methanol/0.1 M NaCl (3:3:1). When separated, the aqueous phase contained roughly 75% of the counts.

G. Thin-Layer Chromatography

The aqueous extract was reduced by lyophilization to approximately 25% of the original volume, and an aliquot (50 μl) was spotted on heat-activated Silica Gel 60 plates. The plates were then developed in either 1-butanol/acetic acid/ether/water (9:6:3:1) or 1-butanol/pyridine/water (4:3:4) and air dried. Radioactivity was located with a Packard model 385 radiochromatogram scanner. The plates were scraped and counted by liquid scintillation.

H. Enzyme Digestion

In the chitinase experiments, the extract was divided into two aliquots: one was treated with 2 mg chitinase (EC 3.2.1.14) in buffer (pH 6.2); the control was treated with buffer only. Both were incubated for 24 hr at 38°C. Volume of the samples was reduced by lyophilization, and a 50-μl aliquot was spotted on a TLC plate.

III. RESULTS AND DISCUSSION

A. Cuticle Spheres

The sclerotized spheres were first observed in the UMBGE-2 cells as round, dark-amber structures with a distinct polygonal surface pattern, apparently the imprint of the secreting cells (Figures 1 and 3). All three samples removed from the freezer showed these structures. Individual secreting cells varied from 50 to 80 μm in diameter, and the cuticle spheres varied from 225 to 275 μm. When treated with concentrated hyamine hydroxide at 80°C for 24 h,

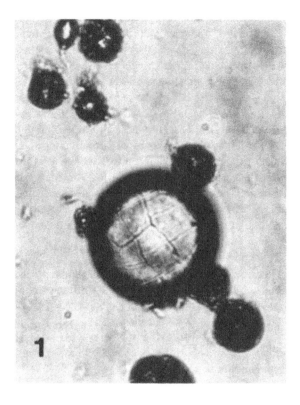

FIGURE 1.   Cuticle sphere from UMBGE-2 cell line derived from
embryos of *Blattella germanica* (22nd passage). Sphere diameter
was 225 μm.

the spheres either increased in diameter (up to 375 μm) or collapsed, but they and a few
individual sclerotized cells remained undigested. Examination of the cultures in which the
cuticle spheres were found revealed numerous cell aggregates about the same size as the
cuticle spheres and delicate membranes from partially formed cuticle spheres. When the
cultures were pelleted, fixed, sectioned, and examined by transmission electron microscopy
(TEM), typical images of insect cuticle were found which showed epi- and endocuticular
structures (Figure 2). Taken together, this evidence leaves little question of the cuticular
nature of these sclerotized spheres.

Alerted to the possible presence of cuticle spheres, we examined frozen cultures of the
MRRL-CH cell lines. Cuticle spheres were found in freshly thawed cultures of both MRRL-
CH and MRRL-CH2 lines (Figure 3). In the former line, samples frozen at passages 23,
29, 34, and 35 were positive. Because the CH and CH2 lines were derived independently[6]
and because the cells from different passages were frozen at different times (as much as 2
years apart), it is clear that the formation of cuticle spheres had occurred on a number of
occasions and is a phenomenon worthy of further investigation.

B. Uptake and Retention of GlcNAc by MRRL-CH Cells

Uptake of GlcNAc remained constant at 11.6×10^4 cpm/h for up to 72 h (Figure 4).
After that, the rate gradually leveled off to 1.4×10^4 cpm/h at 168 h. This rapid and
continued uptake strongly suggested that some mechanism other than simple diffusion was
involved.

When cells dosed with GlcNAc were subsequently washed and incubated in fresh medium,
20% of the incorporated GlcNAc was lost in the first 3 h, and 12% more was lost in the

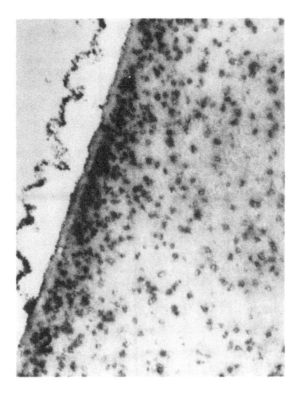

FIGURE 2.   Electron micrograph of a section from a cuticle sphere
from UMBGE cell line. Visible are the delicate epicuticle and the
thicker endocuticle with pore canals. Fixed in 2% OsO_4 and stained
with lead citrate.

remaining 21 h. However, additional experiments confirmed that 68% of the incorporated
GlcNAc was retained against a diffusion gradient for at least 24 h. This again suggested
that something more than simple diffusion was involved.

The nature of the uptake process was further investigated in two experiments during which
the cells were maintained at 2°C in an ice bath. In the first experiment, uptake of GlcNAc
over a 30-min period was measured. Results are given in Table 1. Thus, at 2°C, uptake of
the labeled compound was reduced by 90% from that occurring at 25°C. In the second
experiment, a series of dosage levels from 1 to 15 μl of ^{14}C-labeled GlcNAc was tested at
2°C. Results are given in Table 2. The amount of uptake that occurred at 2°C was directly
dependent upon concentrations of the compound present in the medium. This indicates that
some type of diffusion process was involved.

To determine whether the uptake and retention of GlcNAc required an energy source,
effects of two inhibitors of cellular respiration — oligomycin and antimycin A — were
tested. Results are given in Table 3. The results show that oligomycin produced a 44%
inhibition of uptake of GlcNAc, and antimycin produced a 62.5% inhibition. Uptake is
clearly dependent on cellular respiration.

Studies of substrate specificity of the uptake process were carried out by using GalNAc
and ManNAc as competitive inhibitors of GlcNAc uptake. Results are presented in Table
4. Inhibition of ^{14}C-GlcNAc uptake by GlcNAc was 40%; inhibition by GalNAc, 27%; and
ManNAc was negligible. Thus, the uptake process was selective among the three isomers
of glucosamine and is apparently stereospecific.

In the low temperature experiments, the cells at 2°C still incorporated a small amount of

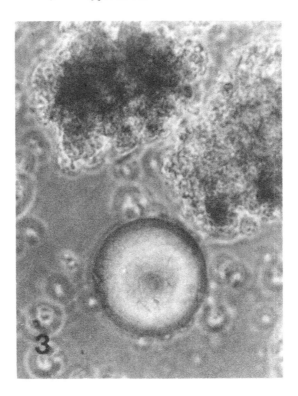

FIGURE 3.   Cuticle sphere from MRRL-CH-34 cell line derived
from *Manduca sexta* (34th passage). Sphere diameter was 272 μm.
Note the delicate pattern of the secreting cells on the surface of the
sphere, similar but smaller than in Figure 1.

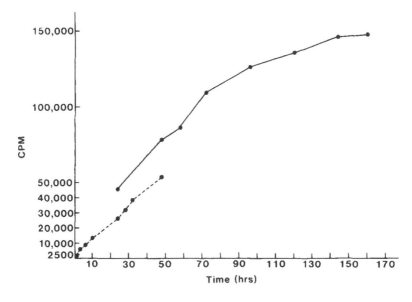

FIGURE 4.   Uptake and retention of [¹⁴C] *N*-acetylglucosamine by MRRL-CH-34 cell line.
Uptake remained constant at 11.6×10^4 cpm/h for up to 72 h.

Table 1
UPTAKE OF ^{14}C-GlcNAc INTO MRRL-CH CELLS AT 2 AND 25°C

Experiment number	Temperature (°C)	Uptake (dpm/million)[a]	Cell population × 10^5 (cells/tube)
1	2	134	9.12
	25	1230	9.12
2	2	147	8.74
	25	1490	8.74

[a]   Mean of three tubes.

Table 2
DEPENDENCE OF GlcNAc UPTAKE AT 2°C BY MRRL-CH CELLS UPON SUBSTRATE CONCENTRATION

Dosage/tube (μl)	Uptake (dpm/million)	Cell population × 10^6 (cells/tube)	Ratio of dosage/tube uptake (dpm/million)
1	124	3.69	1.0:1
3	365	3.36	2.9:1
5	600	3.69	4.8:1
8	900	3.36	7.3:1
10	1250	3.69	10.1:1
15	1810	3.36	14.6:1

Table 3
INHIBITION OF UPTAKE OF ^{14}C-GlcNAc BY OLIGOMYCIN AND ANTIMYCIN A

Experiment number	Incorporation (dpm/million)[a]	Cell population × 10^6 (cells/tube)
3-Oligomycin	317	1.66
Control	558	1.66
4-Oligomycin	301	1.67
Control	551	1.67
5-Antimycin A	224	2.25
Control	517	2.25
6-Antimycin A	128	2.48
Control	401	2.48

[a]   Mean of three tubes.

GlcNAc. Using the cell counter to determine cell volume,[9] we found the intracellular concentration to be 9.6×10^{-6} *M*, just above the extracellular (dosage) concentration of 7.3×10^{-6}. From this it is evident that uptake was due to a facilitated diffusion process. Because maintenance of uptake against a diffusion gradient over a long period requires energy, one or more energy-producing systems must be involved. This conclusion is supported by the partial blockage of uptake by inhibitors of cellular respiration. The 62.5% inhibition of

Table 4
COMPETITIVE INHIBITION OF GlcNAc
UPTAKE BY GlcNAc AND ManNAc

Experiment number	Incorporation (dpm/million)[a]	Cell pupulation × 10[6] (cells/tube)
8-GlcNAc	562	1.56
Control	930	1.56
9-GlcNAc	343	1.41
Control	581	1.41
10-GlcNAc	535	1.07
Control	861	1.07
11-GalNAc	647	1.27
Control	1047[b]	1.27
12-GalNAc	762	1.30
Control	976	1.30
13-GalNAc	354	2.49
Control	406	2.49
14-GalNAc	322	2.28
Control	494	2.28
Control	494	2.28
15-ManNAc	590	3.22
Control	528	3.22
16-ManNAc	604	3.09
Control	481[c]	3.08

[a] Mean of three tubes.
[b] 25 samples only.
[c] Control low due to unequal distribution of cells.

incorporation caused by antimycin A is evidence that cellular respiration is involved since it has been shown that antimycin A inhibits the reoxidation of coenzyme Q in the mitochondrial electron transfer system.[10] The inhibition of uptake by oligomycin (44%) supports this hypothesis. Oligomycin has been shown to be a phosphorylation inhibitor,[11] so its action is directly on the synthesis of the ATP molecule.

Results of the competition studies showed that the uptake process is to some extent stereospecific. Orientation of the *N*-acetyl group distinguishes ManNAc from GlcNAc and GalNAc. In turn, GalNAc differs from GlcNAc at the *N*-acetyl position as well as one or more of the hydroxyl positions.

Taken together, the uptake and retention studies indicated GlcNAc crosses the cell membrane by facilitated diffusion and is held against the gradient by an energy-requiring process that involves conversion to one or more products by one or more stereospecific enzymes. These products may or may not be involved in production of a chitin-containing cuticle.

C. Incorporation of GlcNAc into Intracellular Products

When the pellet of ^{14}C-GlcNAc-dosed cells was washed three times, extracted with ether/water (3:1), and partitioned against water, 75% of the counts were found in the aqueous phase. When the aqueous fraction of the cell extract was reduced by lyophilization, spotted on a Silica Gel TLC plate and developed in a solution of butanol/acetic acid/ether/water (9:6:3:1), virtually all radioactivity remained at the origin ($R_f = 0$) (Figure 5A). When the medium was spiked with ^{14}C-GlcNAc and run, a distinct peak was found at R_f 0.29 (Figure 5B). When the whole aqueous fraction of the cell extract was digested with chitinase for 24 h, the peak at the origin was reduced in size, and a new peak appeared at R_f 0.38 (Figure

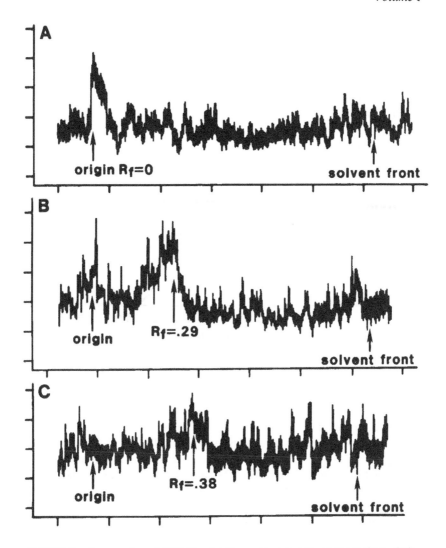

FIGURE 5. Incorporation of [^{14}C] *N*-acetylglucosamine into intracellular products. Cells labeled with [^{14}C] *N*-acetylglucosamine were extracted with ether/water and partitioned against water. The aqueous fraction was chromatographed on a silica gel plate, developed with a solution of butanol/acetic acid/ether/water (9:6:3:1) and scanned by a Packard model 385 scanner. (A) Labeled material remains at origin; (B) same as A but spiked with [^{14}C]-labeled *N*-acetylglucosamine; (C) aqueous fraction digested with chitinase for 24 h. A new peak appears at R_f 0.38.

5C). In a second experiment run under similar conditions, the R_f values were 0.34 for medium in which the cells had been incubated and 0.37 for the chitinase digested cell extract. GlcNAc dissolved in distilled water produced an R_f of 0.27. In a third experiment using a chloroform/methanol extraction procedure followed by chitinase digestion, the digest was spotted and run in the same TLC system as above and then scraped in 5-mm bands and counted by liquid scintillation. This procedure gave slightly different results, with the bulk of the counts remaining at the origin as before but with a distinct peak at R_f 0.27. Clearly, the chloroform/methanol extraction procedure gave a cleaner separation of the products of GlcNAc metabolism. From these experiments, it was apparent that at least a portion of the water-soluble material was made up of higher molecular weight compounds containing GlcNAc linked in the beta-1-4 configuration, which is labile to chitinase.

Incorporation of GlucNAc by MRRL–CH Cells

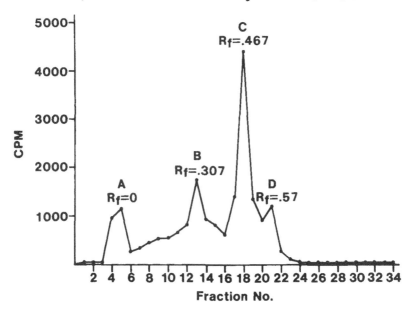

FIGURE 6.   Incorporation of [^{14}C] *N*-acetylglucosamine into MRRL-CH-34 cells. Extracted with chloroform/methanol, spotted on silica gel plates, and developed with a solution of butanol/chloroform/water (4:3:4). The plates were scraped and counted by liquid scintillation.

Because in the original TLC system most of the incorporated GlcNAc remained at the origin, we tried a new solvent system designed specifically for TLC of oligosaccharides (butanol/pyridine/water, 4:3:4). A series of ^{14}C-labeled standards was run in this solvent system, which included GlcNAc, UDPGlcNAc, and chitobiose. TLC of the whole chloroform/methanol extract separated four peaks: A (R_f 0); B (R_f 0.31); C (R_f 0.47); and D (0.57) (Figure 6). When compared with the standards, peak A did not move in this solvent system; peak B had an R_f similar to that of chitobiose; peak C, which contained most of the radioactivity, co-chromatographed with UDPGlcNAc, and peak D co-chromatographed with GlcNAc. When the whole aqueous extract was digested with chitinase for 24 h, peaks A and B were greatly reduced, and a broad single peak with an R_f of 0.4 appeared in which the individual sugars were no longer distinguishable.

Taken together with results obtained with the original solvent system, it is clear that GlcNAc was taken up from the medium and incorporated into at least three products. Most of the activity was found in the peak that co-chromatographed with UDPGlcNAc, which is apparently the first product made as GlcNAc enters the cell and which accounts for the rapid, facilitated uptake of GlcNAc. Peak A, which may be either glycoproteins or a higher polymer of GlcNAc, and peak B are labile to chitinase and, thus, contain the beta-1-4 linkage which is attacked by this enzyme.

IV. CONCLUSIONS

The presence of base-resistant spheres with typical cuticular ultrastructure after freezing in the 35th passage of an insect cell line is sufficient evidence that disruption of embryonic tissues and repeated subculture over a period of time do not necessarily destroy the ability of the cells to produce cuticle-like substances. The absence of this capacity in most cell lines and our inability to produce cuticle spheres at will argue that the cuticle-producing

cells may either be lost after freezing (by overgrowth in subsequent passage) or have lost
the ability to produce one or more components of the mature cuticle. In any case, it is clear
from this preliminary study that much of the physiological machinery for dealing with
GlcNAc remains intact in the MRRL-CH cell line. Whether or not the physiological processes
and metabolic products described so far relate directly to the incorporation into a new cuticle
of GlcNAc liberated during the molting process by the action of chitinase, we do not know.
Many of the details of uptake, metabolism, and redeposition of recycled GlcNAc remain
obscure. Cell culture methodology will be useful in the investigation of these events.

ACKNOWLEDGMENT

We acknowledge with thanks the advice and council of Dr. Harold Klosterman, Department of Biochemistry, North Dakota State University, and the technical assistance of Mr. Jeffrey Balke of the Department of Biochemistry, North Dakota State University, Fargo, ND.

This chapter was published with the approval of the Director of the North Dakota Agricultural Experiment Station as journal article No. 1707. Mention of a company name or proprietary product does not imply endorsement by the U.S. Department of Agriculture.

REFERENCES

1. **Marks, E. P., Leighton, T., and Leighton, F.,** Modes of action of chitin synthesis inhibitors, in *Insecticidal Modes of Action,* Coates, J., Ed., Academic Press, New York, 1982, 281.
2. **Cohen, E. and Casida, J.,** Properties of *Tribolium* gut synthetase, *Pestic. Biochem. Physiol.,* 13, 121, 1980.
3. **Mayer, R. T., Chen, A. C., and DeLoach, J. R.,** Characterization of a chitin synthetase from the stable fly *Stomoxys calcitrans* (L), *Insect Biochem.,* 10, 549, 1980.
4. **Dübendorfer, A., Shields, G., and Sang, J.,** Development and differentiation *in vitro* of *Drosophila* imaginal disk cells from dissociated early embryos, *J. Embryol. Exp. Morphol.,* 33, 487, 1975.
5. **Marks, E. P., Balke, J., and Klosterman, H.,** Evidence for chitin synthesis in an insect cell line, *Arch. Insect Biochem. Physiol.,* 1, 225, 1984.
6. **Eide, P. E., Caldwell, J. M., and Marks, E. P.,** Establishment of two cell lines from embryonic tissue of the tobacco hornworm, *Manduca sexta, In Vitro,* 11, 395, 1975.
7. **Yunker, C., Vaughn, J., and Corey, J.,** Adaptation of an insect cell line (Grace's Antheraea Cells) to a medium free of insect hemolymph, *Science,* 155, 1565, 1967.
8. **Kurtti, T. and Brooks, M.,** Isolation of cell lines from embryos of the cockroach, *Blattella germanica, In Vitro,* 13, 11, 1977.
9. **English, L. E. and Marks, E. P.,** Beta-(2 furyl)-acryloyl phosphate hydrolase activity in insect cell surface membranes, *Biophys. Biochem. Res. Commun.,* 10, 775, 1981.
10. **Pumphrey, A. and Redfern, E.,** The rate of reduction of the endogenous ubiquinone in a heart-muscle preparation, *Biochem. J.,* 72, 2, 1959.
11. **LaRoy, H. A., Johnson, D., and McMurray, W. C.,** Antibiotics as tools for metabolic studies. I. A survey of toxic antibiotics in respiratory phosphorylative and glycolytic systems, *Arch. Biochem. Biophys.,* 78, 587, 1958.

Chapter 6

CHARACTERIZATION OF LEPIDOPTERAN CELL LINES BY ISOELECTRIC FOCUSING AND PHOSPHOGLUCOISOMERASE*

Arthur H. McIntosh and Carlo M. Ignoffo

TABLE OF CONTENTS

---

* Mention of a proprietary product in this paper does not constitute a recommendation for use by the U.S. Department of Agriculture.

I. INTRODUCTION

With the ever increasing number of insect cell lines being established, it is necessary to have adequate means for cell line identification. Within the class Insecta, most cell lines have been established from Diptera (34 species) and Lepidoptera (23 species).[1] Karyologic analysis (distinctive differences in chromosome number and morphology) readily distinguishes dipteran cells from lepidopteran cells. Lepidopteran cells contain both macro- and microchromosomes, and the ploidy is quite variable.[2-4] In contrast, the chromosomal ploidy of dipteran cells is less variable. Dipteran cells have a low diploid number (2n = 6 for mosquitoes; 2n = 8 for *Drosophila* species) and a morphology that is very different from lepidopteran cells.

Some of the more common methods employed in the identification of insect cell lines include: immunologic, karyologic, and enzymatic analyses.[4-12]

In this study, isoenzyme analysis, employing the enzyme phosphoglucoisomerase (PGI) and isoelectric focusing (IEF), was successfully used for both intragenic and intergenic identification of lepidopteran cell lines. Assessment also was made of the effect of cell line passage on isoenzyme pattern and whether or not the isoelectric pattern of the host could be used to identify established cell lines.

II. MATERIALS AND METHODS

Cell lines — The insect cell lines employed in the study were *Spodoptera frugiperda* IPLB-SF21,[13] *Trichoplusia ni* TN-CL1,[14] a clone of TN-368,[15] *Heliothis zea* BCIRL-HZ-AM3,[4] *Heliothis virescens* BCIRL-HV-AM1,[16] *Heliothis armigera* BCRIL-HA-AM1,[17] and *Plutella xylostella*.[18] All cell lines were grown in TC199-MK[19] in T25-cm^2 disposable flasks at 28°C.

Cell extracts and protein determination — Extracts from each cell line were made from 10^7 to 10^8 cells. Confluent cultures were washed three times with calcium- and magnesium-free buffered saline (CMF-PBS)[20] and then removed from the growth vessels with a rubber policeman. Cell suspensions in CMF-PBS were subjected to 1500 lb/in^2 in a French Press (American Ins. Co., Silver Spring, MD). Extracts were clarified by centrifugation (Eppendorf Microfuge; Brinkman Inst., Westbury, NY) at 15,600 × g for 3 min. Protein determinations were performed on clarified extracts employing the Bio-Rad® Assay (Bio-Rad, Richmond, CA). Samples were stored at −70°C.

Isoelectric focusing and phosphoglucoisomerase staining — The methodologies for IEF and PGI staining have been described in a previous report.[4] Commercially readily available polyacrylamide gel plates (PAG) were utilized for results reported herein.

Hemolymph and ovarian tissue — Larvae (3rd to 4th instar) of several *Heliothis* species as well as other noctuids were placed on ice for a few minutes. Then their prolegs were snipped and hemolymph (LH) was collected in tubes held on ice. Ovarian (OVP) tissue was removed from pupae and prepared in the same manner as described for cell extracts. Protein determinations were made on samples, and then the samples were stored at −70°C.

III. RESULTS AND DISCUSSION

Intragenic and intergenic differentiation of six lepidopteran cell lines from four genera were made using IEF and PGI (Figure 1). Intergenic identification of cell lines derived from *Plutella, Spodoptera, Trichoplusia,* and *Heliothis* were differentiated on the basis of major and minor bands. Major bands are readily discernible in Figure 1. Minor bands (less intense in staining), although readily identifiable on the gel, were difficult to resolve photographically. Cells of *P. xylostella* produced one major band (isolectric point [pI] 5.28) and two

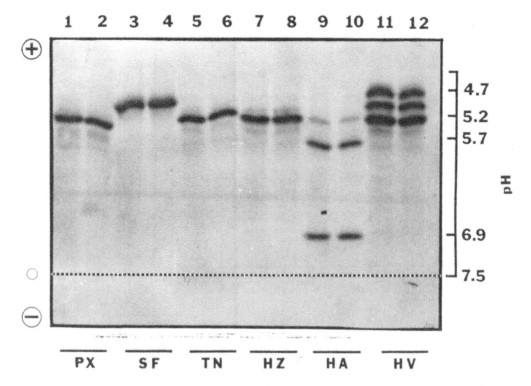

FIGURE 1. Isoenzyme patterns of six lepidopteran cell lines using isoelectric focusing with PGI. From left to right lanes 1 and 2, *P. xylostella* (PX); lanes 3 and 4, *S. frugiperda* (SF); lanes 5 and 6, *T. ni* (TN); lanes 7 and 8, *H. zea* (HZ); lanes 9 and 10, *H. armigera* (HA); lanes 11 and 12, *H. virescens* (HV). Dotted line (— — —) and the open circle (○) indicate sample origin.

minor bands (pI 5.60, 5.82). Major bands and their pI for other cell lines were the following: *S. frugiperda* (pI 5.00), *T. ni* (pI 5.35), *H. zea* (pI 5.35), *H. armigera* (pI 5.35, 5.82, 6.82), and *H. virescens* (pI 4.84, 5.10, 5.35). *T. ni* cells and three cell lines of *Heliothis* shared a common major band (pI 5.35). The *Heliothis* lines, however, were readily distinguished from each other based on pI values and the number of major bands. Cell lines of *T. ni* and *H. zea* are readily distinguished from each other based on their minor bands. *T. ni* cells have two minor bands (at pI 5.54 and 5.82), whereas *H. zea* has five minor bands (at pI 5.28, 5.60, 5.82, 6.68, and 6.82).

Standard errors (SE) of pI means for insect cell lines, larval hemolymph, and ovarian pupae ranged from 0.01 to 0.06. Standard error values are based on duplicate determinations of each sample. A total of ten experiments were performed on cell lines of *H. zea* (SE 0.02), *H. virescens* (SE 0.03, 0.03, 0.06, three major bands), and *H. armigera* (SE 0.03, 0.03, 0.02, three major bands). Seven experiments were conducted on *S. frugiperda* (SE 0.01) and *T. ni* (SE 0.04), and two on *P. xylostella* (SE 0.05). A total of five experiments were conducted on *H. zea* LH (SE 0.02), *H. zea* OVP (SE 0.04), and *H. virescens* LH (SE 0.03, 0.04, 0.05, three major bands). Two experiments were performed on *H. armigera* LH (SE 0.01) and three on *H. virescens* OVP (SE 0.06, 0.05, 0.05, three major bands). Although PGI appears to be the most universal for identifying insect cell lines (it is discriminating and rapid color develops in 5 to 10 min), the procedure also can be applied to other isoenzymes.

Identity is of paramount importance in distinguishing cell lines derived from different genera and from different species within the same genus. Two pertinent questions must be addressed. First, do isoenzyme patterns of cell lines change with *in vitro* passaging? (If major changes occur during passaging, this would make it difficult or impossible to establish

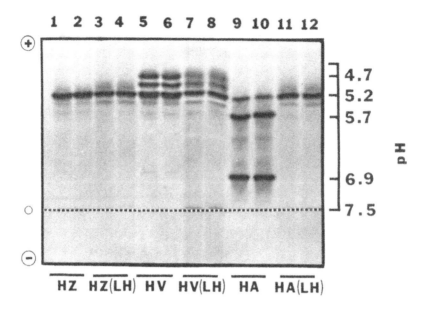

FIGURE 2.   Comparison of isoenzyme patterns between three lepidopteran cell lines and hemolymph from their species of origin by isoelectric focusing with PGI. Lanes 1 and 2, *H. zea* (HZ); lanes 3 and 4, *H. zea* larval hemolymph (HZ[LH]); lanes 5 and 6, *H. virescens* (HV); lanes 7 and 8, *H. virescens* larval hemolymph (HV[LH]); lanes 9 and 10, *H. armigera* (HA); lanes 11 and 12, *H. armigera* larval hemolymph (HA[LH]). Dotted line (— — —) and the open circle (○) indicate sample origin.

cell line identity.) Second, how similar are the isoenzyme patterns of cell lines to the host from which they were derived?

In addressing the first question, a *Heliothis* cell line (BCIRL-HZ-AM3) gave identical patterns at passages 24 and 138 with PGI. The PGI pattern was, thus, stable over a total of 112 passages *in vitro*. Identical isoenzyme patterns of *H. zea* and *H. virescens* cells with their homologous hemolymph were observed (Figure 2). Also, *T. ni* and *S. frugiperda* cell lines gave identical patterns with their hemolymph (data not shown). The exception to this generalization occurred with *H. armigera* (Figure 2) and *S. ornithogalli* (data not shown). The isoenzyme patterns of these two cell lines and their hemolymph were dissimilar. In the case of the *H. armigera* cell line, two major bands (pI 5.82, 6.82) in the isoenzyme pattern were absent in larval hemolymph. The third major band (pI 5.35) was common to both preparations. The reason for the dissimilarity between the isoenzyme patterns of the *H. armigera* and *S. ornithogalli* cell line and its hemolymph is not known. One possible explanation is that the gene(s) encoding these proteins in the cell line is under suppression in the host.

Since the cell lines in this study were derived from ovarian tissue, it was of interest to compare the isoenzyme patterns of ovarian tissue with that of larval hemolymph. The isoenzyme patterns of larval hemolymph and ovarian tissue for each of the two species of *Heliothis* (*H. zea* and *H. virescens*) were identical (Figure 3).

The technique of IEF and isoenzyme analysis is well suited for the identification of insect cell lines. The method is simple, reproducible, and relatively rapid to perform (ca. $1^{1}/_{2}$ h). In addition, PAG, with a wide range of pH values, are readily available commercially. We believe that isoenzyme analysis, employing various electrophoretic techniques (starch gel, cellulose acetate, isoelectric focusing), is currently the most reliable and reproducible method of insect cell line identification.

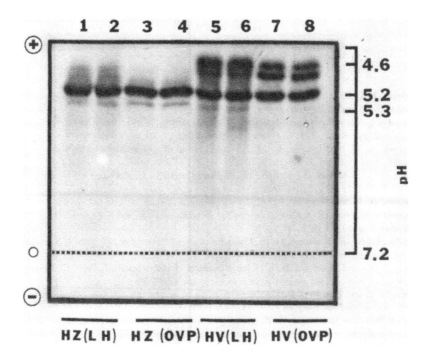

FIGURE 3.   Comparison of isoenzyme patterns between larval hemolymph (LH) and ovarian pupal tissue (OVP) from species of *Heliothis* employing isoelectric focusing with PGI. Lanes 1 and 2, *H. zea* larval hemolymph (HZ[LH]); lanes 3 and 4, *H. zea* ovarian pupal tissue (HZ[OVP]); lanes 5 and 6, *H. virescens* (HV[LH]); lanes 7 and 8, *H. virescens* ovarian pupal tissue (HV[OVP]). Dotted lines (— — —) and the open circle (○) indicate sample origin.

IV. SUMMARY

Isoelectric focusing employing the enzyme PGI was successfully used for intragenic and intergenic identification of six lepidopteran cell lines originating from four genera. pI values of major bands were determined for the following cell lines: *Heliothis zea* (pI 5.35), *Heliothis armigera* (pI 5.35, 5.82, 6.82), *Heliothis virescens* (pI 4.84, 5.10, 5.35), *Plutella xylostella* (pI 5.28), and *Spodoptera frugiperda* (pI 5.00). Although the *Trichoplusia ni* and *H. zea* lines have a major band in common (pI 5.35), they could be readily distinguished from each other based on their minor bands. *T. ni* possesses two minor bands, whereas *H. zea* has five minor bands. The same isoenzyme pattern was obtained after passaging the *H. zea* cell line 112 times. Preparations of hemolymph or ovarian tissue from *H. zea*, *H. virescens*, *T. ni*, and *S. frugiperda* gave identical major band patterns with their homologous cell lines. Only patterns from *H. armigera* and *S. ornithogalli* cell lines were different from their species or origin.

ACKNOWLEDGMENT

We gratefully acknowledge the technical assistance of M. Pappas, P. Andrews, and S. Long.

REFERENCES

1. **Hink, W. F. and Bezanson, D. R.**, Invertebrate cell culture media and cell lines, in *Techniques in Setting Up and Maintenance of Tissue and Cell Cultures*, Kurstak, E., Ed., Elsevier Scientific Publishers, Ireland, 1985, 1.
2. **Thomson, J. A. and Grace, T. D. C.**, Cytological observations on cell strains established in culture from insect ovarian tissue, *Aust. J. Biol. Sci.*, 16, 869, 1963.
3. **Schneider, I.**, Karyology of cells in culture. F. Characteristics of insect cells, in *Tissue Culture Methods and Applications*, Kruse, P. F. and Patterson, M. K., Eds., Academic Press, New York, 1973, 788.
4. **McIntosh, A. H. and Ignoffo, C. M.**, Characterization of five cell lines established from species of *Heliothis*, *Appl. Entomol. Zool.*, 18, 262, 1983.
5. **Greene, A. E., Charney, J., Nichols, W. W., and Corell, L. L.**, Species identity of insect cell lines, *In Vitro*, 7, 313, 1972.
6. **Knudson, D. L. and Buckley, S. M.**, Invertebrate cell culture methods for the study of invertebrate-associated animal viruses, in *Methods in Virology*, Maramorosch, K. and Koprowski, H., Eds., Academic Press, New York, 1977, 323.
7. **Aldridge, C. A. and Knudson, D. L.**, Characterization of invertebrate cell lines. I. Serologic studies of selected lepidopteran lines, *In Vitro*, 16, 384, 1980.
8. **Brown, S. E. and Knudson, D. L.**, Characterization of invertebrate cell lines. III. Isozyme analyses employing cellulose-acetate electrophoresis, *In Vitro*, 16, 829, 1980.
9. **Brown, S. E. and Knudson, D. L.**, Characterization of invertebrate cell lines. IV. Isozyme analyses of dipteran and acarine cell lines, *In Vitro*, 18, 347, 1982.
10. **Herrera, R. J. and Mukherjee, A. B.**, Electrophoretic characterization and comparison of dehydrogenases from eight permanent insect cell lines, *Comp. Biochem. Physiol. B.*, 72, 359, 1982.
11. **Herrera, R. J. and Mukherjee, A. B.**, Electrophoretic characterization and comparison of non-dehydrogenases from ten permanent insect cell lines, *Comp. Biochem. Physiol. B.*, 81, 429, 1985.
12. **Tabachnick, W. J. and Knudson, D. L.**, Characterization of invertebrate cell lines. II. Isozyme analyses employing starch gel electrophoresis, *In Vitro*, 16, 392, 1980.
13. **Vaughn, J. L., Goodwin, R. H., Tompkins, G. J., and McCawley, P.**, Establishment of two cell lines from the insect *Spodoptera frugiperda* (Lepidoptera: Noctuidae), *In Vitro*, 13, 213, 1977.
14. **McIntosh, A. H. and Rechtoris, C.**, Insect cells: colony formation and cloning in agar medium, *In Vitro*, 10, 1, 1974.
15. **Hink, W. F.**, Established insect cell line from the cabbage looper, *Trichoplusia ni*, *Nature (London)*, 226, 466, 1970.
16. **McIntosh, A. H., Andrews, P. A., and Ignoffo, C. M.**, Establishment of a continuous cell line of *Heliothis virescens* (F.) (Lepidoptera: Noctuidae), *In Vitro*, 17, 649, 1981.
17. **McIntosh, A. H., Ignoffo, C. M., Quhou, C., and Pappas, M.**, Establishment of a cell line from *Heliothis armigera* (Hbn.) (Lepidoptera: Noctuidae), *In Vitro*, 19, 589, 1983.
18. **Quhou, C., McIntosh, A. H., and Ignoffo, C. M.**, Establishment of a new cell line from the pupae of *Plutella xylostella*, *J. Cent. China Teacher's Coll.*, 3, 99, 1983.
19. **McIntosh, A. H., Maramorosch, K., and Rechtoris, C.**, Adaptation of an insect cell line *(Agallia constricta)* in a mammalian cell culture medium, *In Vitro*, 8, 375, 1973.
20. **Merchant, D. L., Kahn, R. H., and Murphy, W. H.**, *Handbook of Cell and Organ Culture*, Burgess, Minneapolis, 1964, 240.

Chapter 7

ISOZYME CHARACTERIZATION OF 8 HYMENOPTERAN AND 20 LEPIDOPTERAN CELL LINES

G. T. Harvey and S. S. Sohi

TABLE OF CONTENTS

I. INTRODUCTION

Several continuous cell lines have been developed at the Forest Pest Management Institute (FPMI) from the tissues of major forest insect pests to facilitate the *in vitro* investigations of insect pathogens. These cell lines originated from the spruce budworm (*Choristoneura fumiferana*), the forest tent caterpillar (*Malacosoma disstria*), the red-headed pine sawfly (*Neodiprion lecontei*), and the white-marked tussock moth (*Orgyia leucostigma*). Isozyme analyses were performed to confirm the identity of the new cell lines. Results of these analyses are presented here.

II. MATERIALS AND METHODS

A. Cell Lines and Culture Media

Twenty-eight continuous insect cell lines originating from four lepidopterous and one hymenopterous species (Table 1) were used in this work. All the cell lines except one were developed at FPMI. They were grown in Grace's insect tissue culture medium[1] supplemented with 0.25% w/v tryptose broth and 10 to 15% v/v fetal bovine serum (heat-treated at 56°C for 30 min). *Trichoplusia ni* cells, TN-368,[2] were grown in Hink's TNM-FH culture medium[2] without whole chicken egg ultrafiltrate.

Three *C. fumiferana* cell lines (FPMI-CF-50, -60, and -70) and the eight *N. lecontei* lines (Table 1), all of which grew attached to the culture flasks, were released by trypsinization.[3] All other cells (Table 1) grew freely suspended in the medium or lightly attached to the culture flasks. They were dislodged with a rubber scraper, centrifuged, rinsed three times with balanced salt solution, and the cell pellets were stored at −80°C prior to use.

B. Insects

Last-instar larvae were used to characterize the isozymes in the five insect species. Larvae of *C. fumiferana* were from field collections. Larvae of the other four species were from laboratory stocks maintained at FPMI: *N. lecontei* on red pine foliage and the other three species on artificial diet.[4] All larvae were stored at −80°C for short periods.

C. Electrophoresis

Freshly thawed, decapitated larvae were homogenized individually in 0.25 ml Tris-citrate buffer (pH 7.1).[3] After centrifugation (8000 × g), the supernatant was used for electrophoresis. Cell pellets were suspended in buffer and centrifuged to produce a cell-free extract for electrophoresis. During preparation, extracts were kept at 0°C.

Insects and cells were analyzed for nine enzymes: aspartate aminotransferase, esterase, isocitrate dehydrogenase, lactate dehydrogenase, leucine aminopeptidase, malate dehydrogenase, malic enzyme, phosphoglucoisomerase, and phosphoglucomutase. Methods of electrophoresis using horizontal starch gels followed Ayala et al.[5] and Hudson and Lefkovitch.[6] Isozyme identification in *C. fumiferana* was based on extensive studies including genetic tests.[7,8] Sufficient numbers of the other four insects were tested to determine the principal bands for each locus and to establish relative frequencies of the allozymes.

III. RESULTS AND DISCUSSION

A. Isozymes in Insects

Nine enzymes represented by as many as 17 loci were present in all five insect species. For most enzyme systems, band locations were characteristic for each insect species and permitted clear differentiation among them. Examples of these differences in three enzymes are presented in Table 2.

Table 1
INSECT CELL LINES EXAMINED BY ISOZYME ANALYSIS

Species	Explant	Cell line designation	Passage	Ref.
C. fumiferana	Neonate larvae	IPRI-CF-1	88	13
		IPRI-CF-5	108	13
		IPRI-CF-6	86	13
		IPRI-CF-8	93	13
		IPRI-CF-10	65	13
		IPRI-CF-12	90	13
		IPRI-CF-16	93	13
		IPRI-CF-124	236	14
		IPRI-CF-124-2	81	14
	Pupal ovaries	FPMI-CF-50	29	13
		FPMI-CF-60	25	13
		FPMI-CF-70	12	13
M. disstria	Larval hemocytes	IPRI-MD-66	174	15
T. ni	Adult ovaries	TN-368	50[a]	2
O. leucostigma	Neonate larvae	IPRI-OL-4	54	13
		IPRI-OL-7	47	13
		IPRI-OL-9	75	13
		IPRI-OL-11	55	13
		IPRI-OL-12	100	13
		IPRI-OL-13	253	13
N. lecontei	Embryos	FPMI-NL-2	35	16
		FPMI-NL-4	32	16
		FPMI-NL-10	31	16
		FPMI-NL-18	51	16
		FPMI-NL-21	64	16
		FPMI-NL-22	44	16
		FPMI-NL-28	96	16
		FPMI-NL-32	23	13

[a] This represents the number of serial passages TN-368 cells have undergone in our laboratory. Cells had already been subcultured over 300 times by Dr. W. F. Hink's laboratory.

A single phosphoglucomutase (PGM) locus was detectable in all species. PGM was monomorphic in *N. lecontei* and polymorphic in the other four species, but no band positions were common to any two species (Table 2). Aspartate aminotransferase (AAT) is produced by two loci in *Choristoneura* species;[8] the anodal locus, AAT-1, is dimeric and sex linked.[9] AAT had one monomorphic locus in the other four species. Most of the AAT bands were quite distinct among the five species (Table 2). Isocitrate dehydrogenase (IDH) is produced by two anodal loci in *Choristoneura* and in *Malacosoma* but only by a single locus in the other three species. In *Orgyia* most insects had two bands in IDH-1 (Table 2). IDH-1 band locations were quite different for the five species. These three loci (PGM, AAT-1, and IDH-1) were adequate to give clear and unequivocal differentiation among the five insect species. There were differences at the remaining loci, but bands were not unique among the insects and, therefore, not useful for differentiating the species.

B. Isozymes in Cell Lines
Not all enzymes present in the insects were detected in the cell lines, but many showed considerable activity. Isozyme bands in all but two cell lines were readily identified with bands in the insect species from which the cultures originated (data are not shown but agreed with Table 2). Bands at the three loci named were adequate to confirm the species identity of 26 of the 28 cell lines tested. The other loci helped confirm cell line identification and permitted differentiation of otherwise identical cell lines.

Table 2
BAND POSITIONS (Rf VALUES) AND FREQUENCIES OF ALLELES OF THREE ENZYMES IN FIVE INSECT SPECIES

Allele Frequency

R_f values	C. fumiferana (N = 2000)	O. leucostigma (N = 44)	M. disstria (N = 11)	N. lecontei (N = 59)	I. ni (N = 17)
Aspartate Aminotransferase — 1					
0.250	0.06				
0.330	0.30				
0.410	0.64	**			
0.420					1.00
0.470				1.00	
0.490	0.002		1.00		
0.510		1.00			
Isocitrate Dehydrogenase — 1					
0.092				1.00	
0.137				1.00	
0.170		0.04			
0.199		0.91			
0.205	0.02				
0.262		1.00			
0.269	0.002				
0.302			1.00		
0.320	0.93	**			
0.352	0.04				
Phosphoglucomutase					
0.146				1.00	
0.250			0.05		
0.260			0.36		
0.274					0.09
0.300			0.59		
0.310					0.91
0.370	0.04				
0.410	0.88	**			
0.430		0.38			
0.440	0.09				
0.470		0.58			
0.510		0.05			

Note: ** indicates positions of bands in OL-7 and OL-11.

All 12 *C. fumiferana* (CF) cell lines showed significant activity at most loci. However, there was variation in some enzymes, and no two CF cell lines were identical at all loci. Nevertheless, all of the isozymes found in all these cell lines were readily identified with isozymes in the insect, thus confirming their species identity.

Isozymes in the eight *N. lecontei* cell lines showed very little variability, as was also noticed for sawfly larvae. All enzymes detected conformed to those present in the insect. This confirmed the *N. lecontei* origin of these lines. Isozymes in the cells of *M. disstria* and *T. ni* were characteristic of the respective species, thus confirming their identity.

Activity of all major enzymes was readily detected in the six cell lines of *O. leucostigma* (OL), and band positions in all but two OL cell lines conformed to band positions in this species. Thus, the identity of all but two OL cell lines was confirmed. Band positions for OL-7 and OL-11 differed from those in *O. leucostigma* larvae but were in closest agreement with bands in *C. fumiferana* (Table 2). Also, in serological and virological studies,[10,11] these two cell lines behaved like CF lines rather than OL lines. We have, therefore, concluded that these OL cell lines were probably contaminated with or replaced by *C. fumiferana* cells. Further analysis pointed to a specific cell line, CF-124, as the probable source of the contamination.[3]

The incorrect identity of two cell lines may have occurred through cross contamination and/or mislabeling of cultures in our laboratory. These results demonstrate the need for constant surveillance and periodic testing to ensure correct identity of cell cultures. This work also supports the conclusions of Tabachnick and Knudson[12] about the reliability and usefulness of isozyme analysis for the identification and monitoring of invertebrate cell lines.

ACKNOWLEDGMENTS

We express sincere appreciation to Mrs. P. Roden, Mrs. B. J. Cook, and Mr. G. F. Caputo for their excellent technical assistance. We also thank the National Research Council for permission to use some previously published material.[3]

REFERENCES

1. **Grace, T. D. C.,** Establishment of four strains of cells from insect tissues grown *in vitro, Nature (London),* 195, 788, 1962.
2. **Hink, W. F.,** Established insect cell lines from the cabbage looper, *Trichoplusia ni, Nature (London),* 226, 466, 1970.
3. **Harvey, G. T. and Sohi, S. S.,** Isozyme characterization of 28 cell lines from five insect species, *Can. J. Zool.,* 63, 2270, 1985.
4. **Grisdale, D.,** An improved laboratory method for rearing large numbers of spruce budworm, *Choristoneura fumiferana.* (Lepidoptera: Tortricidae), *Can. Entomol.,* 102, 1111, 1970.
5. **Ayala, F. J., Powell, J. R., Tracey, M. L., Mourão, C. A., and Pérez-Salas, S.,** Enzyme variability in the *Drosophila willistoni* group. IV. Genic variation in natural populations of *Drosophila willistoni, Genetics,* 70, 113, 1972.
6. **Hudson, A. and Lefkovitch, L. P.,** Allozyme variation in four Ontario populations of *Xestia adela* and *Xestia dolosa* and in a British population of *Xestia c-nigrum* (Lepidoptera: Noctuidae), *Ann. Entomol. Soc. Am.,* 75, 250, 1982.
7. **Harvey, G. T.,** unpublished data, 1987.
8. **Stock, M. W. and Castrovillo, P. J.,** Genetic relationships among representative populations of five *Choristoneura* species: *C. occidentalis, C. retiniana, C. biennis, C. lambertiana* and *C. fumiferana,* (Lepidoptera: Tortricidae), *Can. Entomol.,* 113, 857, 1981.
9. **May, B., Leonard, D. E., and Vadas, R. L.,** Electophoretic variation and sex linkage in spruce budworm, *J. Hered.,* 68, 355, 1977.
10. **Sohi, S. S. and Krywienczyk, J.,** unpublished data, 1987.
11. **Sohi, S. S., Caputo, G. F., and Cook, B. J.,** Replication and cross-infectivity of five baculoviruses in cell cultures, unpublished report, 1988.
12. **Tabachnick, W. J. and Knudson, D. L.,** Characterization of invertebrate cell lines. II. Isozyme analyses employing starch gel electrophoresis, *In Vitro,* 16, 392, 1980.
13. **Sohi, S. S.,** unpublished data, 1987.
14. **Sohi, S. S.,** In vitro cultivation of larval tissues of *Choristoneura fumiferana* (Clemens) (Lepidoptera: Tortricidae), in *Proc. 3rd Int. Colloq. Invertebr. Tissue Cult.,* Smolenice, Czechoslovakia, (1971), 75, 1973.
15. **Sohi, S. S.,** In vitro culture of hemocytes of *Malacosoma disstria* Hübner (Lepidoptera: Lasiocampidae), *Can. J. Zool.,* 49, 1355, 1971.
16. **Sohi, S. S. and Ennis, T. J.,** Chromosomal characterization of cell lines of *Neodiprion lecontei* (Hymenoptera: Diprionidae), *Proc. Entomol. Soc. Ont.,* 112, 45, 1981.

Chapter 8

ELECTRON MICROSCOPIC STUDIES ON *IN VITRO* DIFFERENTIATED CELLS FROM *DROSOPHILA* EMBRYOS

Y. Kuroda and Y. Shimada

TABLE OF CONTENTS

I. INTRODUCTION

During embryonic development a specific gene at a specific site on the chromosomes expresses its action at one specific time in certain cells and tissues. Many genes may participate in differentiation of each type of cells and in the formation of various tissues and organs in early development of embryos. It is important to study the specific time and the specific site of the expression of these genes to understand the genetic control of the development.

In preceding work by the authors on the cultivation of embryonic cells from X-linked recessive lethal mutant strains, specific defects were observed in differentiation and function in certain types of cells.[1-3] Addition of an extract of unfertilized wild-type eggs to the culture medium resulted in repair of these defects. Undifferentiated cells from wild-type gastrula differentiated into muscle cells, epithelial cells, fibroblastic cells, and nerve cells, which showed their characteristic features and function under a light microscope.

In the present paper, the ultrastructural features of these differentiated cells were examined under an electron microscope.

II. MATERIALS AND METHODS

Cultivation of embryonic cells — Newly laid eggs were collected from a wild-type strain (Oregon-R) of *Drosophila melanogaster*. The eggs were dechorionated by treatment with 3% sodium hypochloride solution for 6 min and then washed with distilled water. The developmental stages of dechorionated eggs were readily determined through the transparent vitelline membrane under a binocular microscope. Embryos at the postgastrula stage were selected, sterilized in 70% ethyl alcohol for 10 min, and washed three times with sterile physiological salt solution (0.7 g NaCl, 0.02 g KCl, 0.002 g $CaCl_2 \cdot 2H_2O$, 0.01 g $MgCl_2 \cdot 6H_2O$, 0.005 g $NaHCO_3$, 0.002 g $NaH_2PO_4 \cdot 2H_2O$, and 0.08 g glucose in 100 ml of distilled water). One hundred embryos were transferred to culture medium in a hollow slide and torn into small fragments with a pair of fine needles under a binocular microscope. Then, small torn fragments of tissues and cells were transferred into culture medium on a carbon-coated coverslip in a T-5 culture flask. The flasks were closed with a stopper and incubated at 28°C.

Culture medium — The culture medium consisted of K-17 medium[1,4] supplemented with 0.1 mg/ml fetuin (Grand Island Biol. Co., Deutsch Method) and 15% fetal bovine serum (Microbiol. Assoc. Inc.).

Electron microscopic observation — After the flasks were incubated for several days, the coverslips on which tissue fragments and cells were attached were taken out of the flasks, fixed in 2.5% glutaraldehyde in 0.1 M phosphate buffer at pH 7.3, and then postfixed in 1% OsO_4 in the same buffer. After dehydration in an ascending ethanol series, the fixed materials were embedded in Epon 812 and split off the coverslips. Thin sections were cut with an LKB ultrotome and stained with uranyl acetate and lead citrate. They were examined with a Hitachi® H-12 or JOEL 100C electron microscope operated at 75 kV.

III. RESULTS

A. Noncharacterized Cells

When cells from embryos at the postgastrula stage were cultured in T-5 flasks for several days, many characteristic cells differentiated. Muscle cells, epithelial cells, and nerve cells were easily identified in their morphology and functions in culture. However, some cells which had no characteristic differentiated morphology and functions were observed. Figure 1 shows such noncharacterized cells which formed clusters in culture. These cells were

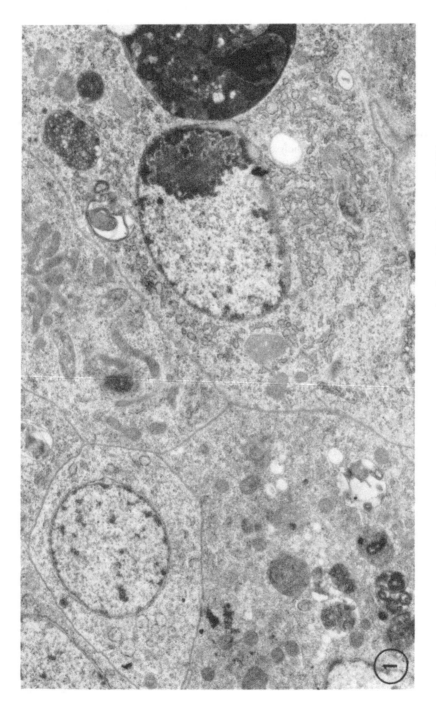

FIGURE 1.   Noncharacterized cells which formed a cluster in culture. (Magnification ×38,000.)

fibroblastic in shape and showed rapid extension and contraction of the cytoplasm. They migrated rapidly by extension of broad or narrow cytoplasmic processes, occasionally reaching a speed of one cell length per hour (about 30 μm/h).

They were arranged closely to each other, but no junctional specializations were found between apposed cells. Their cytoplasm contained free ribosomes, rod-shaped mitochondria, and granular endoplasmic reticulum, but the cytoplasm was devoid of structures suggestive of a particular cell type. These cells sometimes contained electron-opaque, membrane-bound granules, which were similar in appearance to the autophagic vacuole variety of lysosome. In the center of the cell, a single, rounded nucleus was found. It possessed a conspicuous nucleolus and dense clumps of opaque material beneath the nuclear membrane.

Other noncharacterized cells, which were separated by wide intercellular spaces, were seen after cultivation for 6 d (Figure 2). These cells were dark in appearance because of the presence of numerous ribosomes in their cytoplasm. Their nuclei were irregular in shape, contained one or two nucleoli, and their nucleoplasm was slightly darker than that of the previous type of cells. Many slender cytoplasmic processes were found in wide, intercellular spaces. These cells also showed little structural evidence of differentiation into particular cell types. There is a possibility that those two types of cells were fibroblastic, and they remained in an undifferentiated state in culture.

B. Muscle Cells

After cultivation for several days, spindle-shaped muscle cells came out from the cut ends of tissue fragments. They extended their cytoplasm over the glass surface and migrated actively. They were 50 to 100 μm in length and had an ovoid nucleus. They grew, made contact with each other, and fused to form syncytial complexes, some containing more than ten nuclei.

Under the electron microscope, the unfused muscle cells elongated and contained many incompletely organized myofibrils in the cytoplasm after cultivation for 6 d. In some of these muscle cells, thick and thin myofilaments were aggregated parallel or nearly parallel, and these aggregates of filaments exhibited Z-band density (Figure 3a).

Other muscle cells made contact with each other and fused to form syncytial complexes, some containing more than ten nuclei. In the syncytium of fused muscle cells, well-developed myofibrils were seen: myofibrils increased in number and diameter, and the characteristic M and Z lines in the A and I bands of sarcomeres were clearly discernible (Figure 3b). Nuclei were arranged longitudinally in the central axis of the cells. This suggests that these multinucleated cells were formed by fusion of mononucleated cells. These cells possessing multinuclei and sarcomere structures were clearly identified as myotomes differentiated *in vitro*.

C. Nerve Cells

Some nerve fibers were found to extend from fragments of nerve tissue. They were fine and thread-like and began to extend after a few hours of cultivation. After several days, the nerve fibers formed branches in many directions from their thicker, main axes. Under the electron microscope, cells which possessed long cable-like extensions tended to form aggregates (Figure 4a). The cell bodies were generally rounded, and their perikaryon contained a large number of ribosomes associated with flattened cisternae of the endoplasmic reticulum as well as those in the cytoplasmic matrix (Figure 4b). Cytoplasmic extensions which radiated out from the aggregates formed bundles (Figure 4a and c). Within the cytoplasmic extensions microtubules were seen running along their long axis. Near the terminal of processes, mitochondria and an accumulation of dense-cored vesicles were found. We interpret these cells as neurons and the processes as axons.

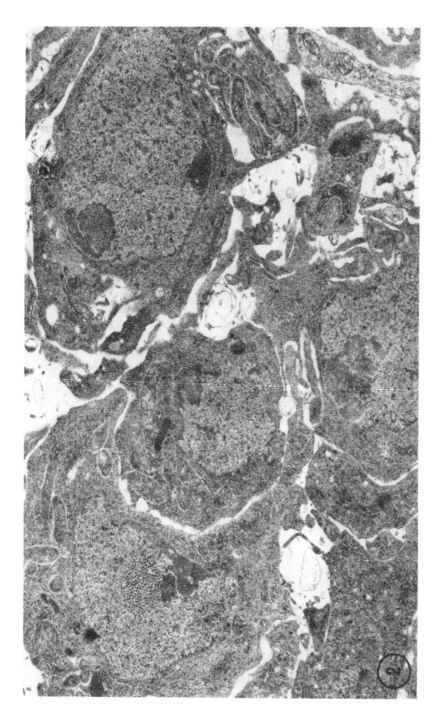

FIGURE 2.   Noncharacterized cells which were separated by wide intercellular spaces. (Magnification ×15,000.)

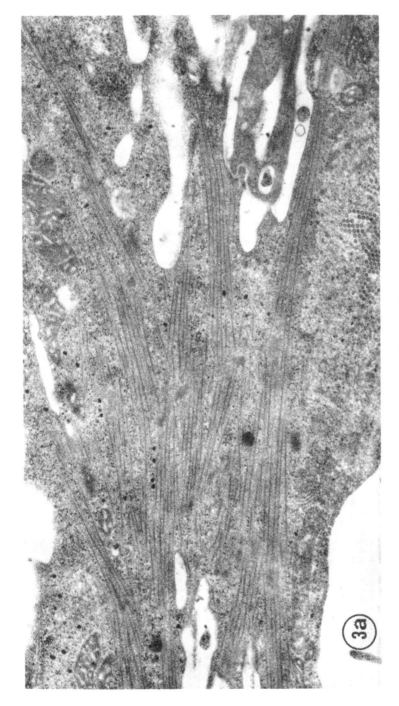

FIGURE 3.   Muscle cells differentiated in culture. (a) An unfused muscle cell. (Magnification ×21,000.) (b) Fused muscle cells which formed a syncytial complex. (Magnification ×22,000.)

FIGURE 3b

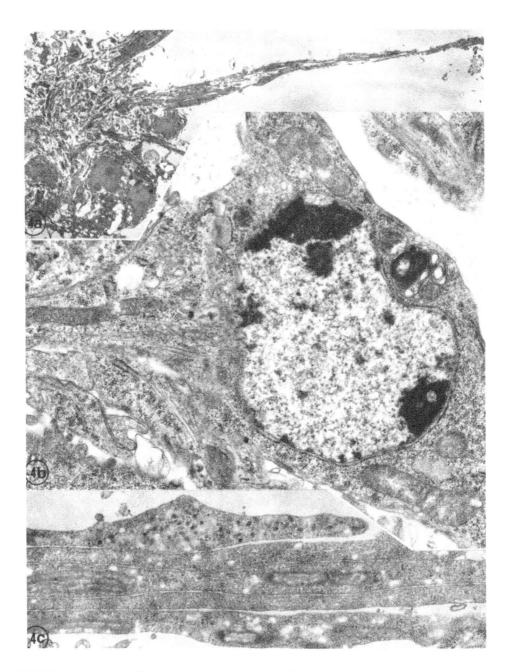

FIGURE 4.   Nerve cells differentiated in culture. (a) Nerve cells which possessed long cable-like extensions. (Magnification ×3700.) (b) A nerve cell body. (Magnification ×9800.) (c) Cytoplasmic extension of a nerve cell. (Magnification ×4000.)

D. Epithelial Cells

Flat, polygonal epithelial cells were clearly distinguished from the other types of cells described above. They were released from the peripheral regions of some tissue fragments after several hours of cultivation. They were about 5 μm in diameter and had a relatively large nucleus with a prominent nucleolus. At the early period of cultivation, their cytoplasm was clear with a few dark granules which gradually increased in size and number during cultivation. After cultivation for 24 h, these cells formed a continuous monolayer sheet.

Under the electron microscope, these epithelial cells possessed a great number of cylindrical processes, or microvilli, at their free, apical surfaces (Figure 5a). These processes were about 0.1 to 0.2 μm in diameter and reached a length of several micrometers (Figure 5b). The cytoplasm within the villi was composed of fine filamentous material which showed a preferred orientation along the axis of the microvilli. The cytoplasm was richly supplied with cisternae of rough endoplasmic reticulum. The nucleus was located in the mid-basal part of the cell. Round- or rod-shaped mitochondria were scattered within the cytoplasm. Lateral surfaces of cells were linked in some places by fascial adherents with symmetrical aggregations of electron-opaque material toward the membrane. In other places, the membrane of one cell showed a finger-like projection toward another cell. These cells were similar in appearance to those of the intestinal epithelium.

The cytoplasm of another type of columnar cell was electron-dense because of the presence of a large number of ribosomes and glycogen particles (Figure 6a). These cells were linked by adherent junctions at their juxtaluminal lateral surfaces (Figure 6b).

E. Round Cells

Round cells were seen to join by adherent junctions (Figures 7a and b). These cells contained numerous vacuoles and dense bodies. Cytoplasmic vesicles containing caveolae were also found. Presumably they represent invaginated and, subsequently, phagocytosed, apical, surface cell membranes with numerous short processes. Rough endoplasmic reticulum formed a lamellated body. Mitochondria of various sizes were scattered within the cytoplasm. Cells of this kind were unable to identify tissue types. However, because of the presence of junctions and dense vesicles, they appeared to be epithelial and secretory in nature. Aggregates of virus-like particles, occasionally arranged in crystalloid, were found in the nucleus and cytoplasm of almost all kinds of cells.

IV. DISCUSSION

Many attempts have been made to obtain the primary cultures of *Drosophila*.[5-9] Schneider[10] initiated cultures with fragments of half or one third embryos, but in all other works cultivation was initiated by dissociating cells by mechanical homogenization.

Shields et al.[11] also initiated cultures by homogenizing embryos in the culture medium. They reported that the subsequent development of the cultures was much improved by mechanical dissociation compared with trypsinization. We also tried to dissociate embryonic cells by mechanical homogenization in earlier experiments. Single cells obtained in this way, however, became attached to the glass surface. They grew for a few days, but most of them then degenerated and became detached from the glass surface. When the cultures were initiated with embryonic tissue fragments, many cells migrated out from the fragments and differentiated to characteristic types of cells — muscle cells, nerve cells, and epithelial cells, as described in the present paper.

Shields et al.[11] found that the differentiating cells which appeared in the primary cultures of $6\frac{1}{2}$ to $8\frac{1}{2}$ *Drosophila* embryos fell into two main groups. The first group involved cells that differentiated very quickly into their final histological form. A second group involved cells that remained quiescent for a week or more, then resumed multiplication, and only slowly differentiated into their final form. The first group included nerve, muscle, fat body, and chitin-secreting cells as well as the first stage of development of the macrophage-like cells. The second group included a second stage of development of the macrophage-like cells along with tracheal and imaginal disk cells and a number of unidentified fibroblastic and epithelial cells.

Kuroda[1,2] found that undifferentiated cells, from embryos at the stage of sac-like midgut (12 h), differentiated into several larval types of cells, muscle cells, epithelial cells, fibro-

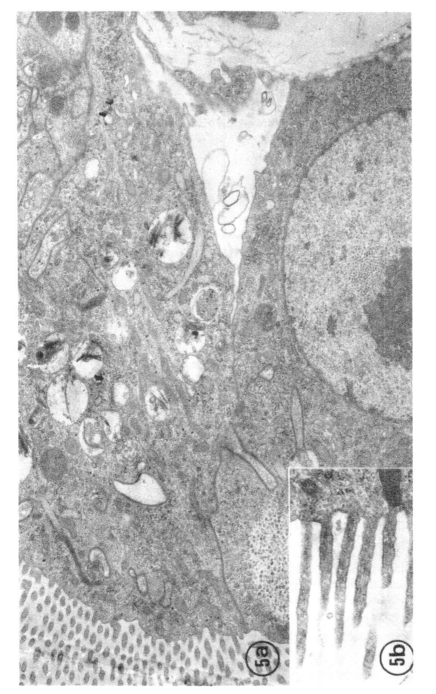

FIGURE 5. Epithelial cells differentiated in culture. (a) An epithelial cell with cylindrical processes or microvilli. (Magnification ×38,000.) (b) Larger magnification of microvilli. (Magnification ×65,000.)

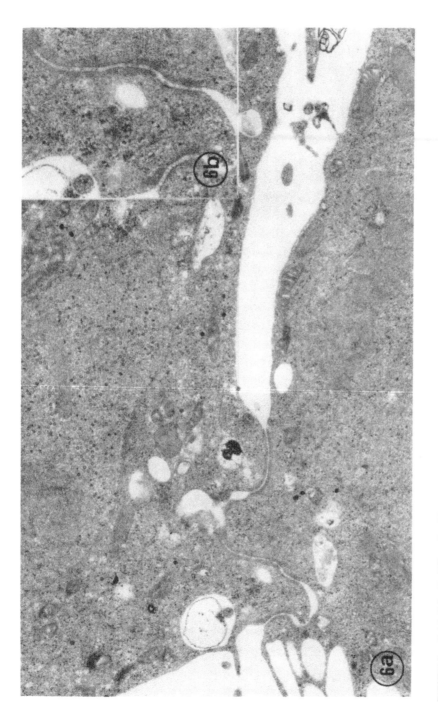

FIGURE 6. Epithelial cells differentiated in culture. (a) An epithelial cell with a large number of ribosomes and glycogen particles. (Magnification ×24,000.) (b) Adherent junctions at the juxtaluminal lateral surface. (Magnification ×62,000.)

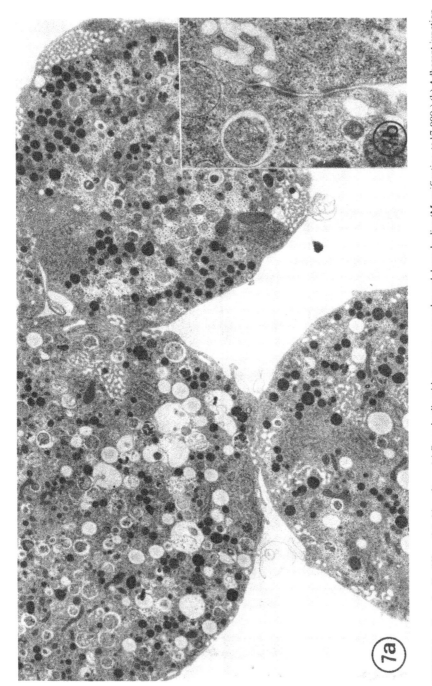

FIGURE 7.   Round cells differentiated in culture. (a) Round cells with numerous vacuoles and dense bodies (Magnification × 17,000.) (b) Adherent junction between two round cells. (Magnification × 73,000.)

blastic cells, and small cells when cultured in the absence of ecdysterone in medium. In the presence of ecdysterone, Kuroda[12,13] found some characteristic adult structures appeared in cultures of undifferentiated embryonic cells.

The former larval types of cells may correspond to the first group of cells by Shields et al.,[10] and the latter cells of adult structures may correspond to the second group of cells. Thus, we have established a culture system of *Drosophila* embryonic cells in which undifferentiated cells can preferentially differentiate into adult-type structures or larval-type cells by adding ecdysterone or not to culture media.

By using this system, Kuroda[3,14] revealed the tissue and time specificities of the action of some sex-linked recessive lethal genes. In these studies, the characteristic features and functions of differentiated cells were identified by a light microscope.

In the present study, we observed some of these differentiated cells by an electron microscope. Muscle cells, nerve cells, and epithelial cells developed in the absence of ecdysterone were larval types of cells, which we could clearly recognize by their characteristic ultrastructures. More detailed analyses may be possible on the tissue and time specificities of the action of lethal genes in normal development by the ultrastructural and molecular study of embryonic cells cultured *in vitro* in this system.

ACKNOWLEDGMENTS

Work carried out by the authors was supported in part by a grant-in-aid from the Ministry of Education, Science, and Culture of Japan. The authors wish to thank Drs. S. Masuko and Y. Isobe for their help in electron microscopy. This is contribution No. 1948 from the National Institute of Genetics, Misima, Japan.

REFERENCES

1. **Kuroda, Y.,** Studies on *Drosophila* embryonic cells *in vitro.* I. Characteristics of cell types in culture, *Dev. Growth Differ.,* 16, 55, 1974.
2. **Kuroda, Y.,** *In vitro* activity of cells from genetically lethal embryos of *Drosophila, Nature (London),* 252, 40, 1974.
3. **Kuroda, Y.,** Studies on *Drosophila* embryonic cells *in vitro.* II. Tissue- and time-specificity of a lethal gene, *deep orange, Dev. Growth Differ.,* 19, 57, 1977.
4. **Kuroda, Y.,** Spermatogenesis in pharate adult testes of *Drosophila* in tissue cultures without ecdysones, *J. Insect Physiol.,* 20, 637, 1974.
5. **Horikawa, M. and Fox, A. S.,** Culture of embryonic cells of *Drosophila melanogaster in vitro, Science,* 145, 1437, 1964.
6. **Gvosdiov, V. A. and Kakpakov, V. T.,** Culture of embryonic cells of *Drosophila melanogaster in vitro* (in Russian), *Genetika,* 2, 129, 1968.
7. **Seecof, R. L. and Unanue, R. L.,** Differentiation of embryonic *Drosophila* cells *in vitro, Exp. Cell Res.,* 50, 654, 1968.
8. **Shields, G. and Sang, J. H.,** Characteristics of five cell types appearing during *in vitro* culture of embryonic material from *Drosophila melanogaster, J. Embryol. Exp. Morphol.,* 23, 53, 1970.
9. **Echalier, G. and Ohanessian, A.,** *In vitro* culture of *Drosophila melanogaster* embryonic cells, *In Vitro,* 6, 162, 1970.
10. **Schneider, I.,** Cell lines derived from late embryonic stages of *Drosophila melanogaster, J. Embryol. Exp. Morphol.,* 27, 353, 1972.
11. **Shields, G., Dübendorfer, A., and Sang, J. H.,** Differentiation *in vitro* of larval cell types from early embryonic cells of *Drosophila melanogaster, J. Embryol. Exp. Morphol.,* 33, 159, 1975.
12. **Kuroda, Y.,** Differentiation of adult structures from *Drosophila* embryonic cells in culture, in *The Ultrastructure and Functioning of Insect Cells,* Akai, H., King, R. C., and Morohoshi, S., Eds., Society for Insect Cells, Japan, 1982, 91.
13. **Kuroda, Y.,** Differentiation of adult structures in cultures of embryonic tissues from *Drosophila melanogaster,* in *Techniques in in Vitro Invertebrate Hormones and Genes,* C215, Elsevier, County Clare, Ireland, 1986, 1.

Endocrinology in Invertebrate Tissue Culture

Chapter 9

RELATIONSHIPS BETWEEN ECDYSTERONE-INDUCED CELLULAR DIFFERENTIATION AND AEROBIOSIS IN AN *IN VITRO* DROSOPHILA CELL SYSTEM

A. M. Courgeon, M. Ropp, E. Rollet, J. Becker, C. Maisonhaute, G. Echalier, and M. Best-Belpomme

TABLE OF CONTENTS

I. INTRODUCTION

For several years, we have shown that *Drosophila* cells of several clones derived from *in vitro*-established lines were target cells for ecdysterone, a steroid hormone controlling differentiation and development in insects. Those clones, which are called sensitive or responsive clones, were shown to have specific receptors for the hormone, contrary to resistant clones.[1] The hormone-sensitive cells display a wide range of responses including arrest of divisions, morphological modifications, induction of enzymes which are known to vary during development, and induction of proteins immunologically related to those found in the adults. (For a review, see Peronnet et al.[2]) These responses may be considered the first steps of a cellular differentiation induced by the hormone. We attempted to study the energetic cost of this differentiation and the relationships between the cellular differentiation and the aerobic metabolism of the cells.

II. MATERIALS AND METHODS

The diploid clones FC and 89K, derived from line Kc, were grown in monolayers in plastic flasks with 3 ml of D22 medium plus 5% fetal calf serum, at 23°C.[3] The thickness of the liquid was 1 mm. The gas phase was air (1 atm, 760 mmHg). The maximal dissolved O_2 concentration was calculated to be 250 μM (O_2 partial pressure: 150 mmHg, Bensen coefficient 0.028). Oxygen consumption was measured with a Clark electrode. The rate of oxygen consumption was expressed as a function of the O_2 concentration and the results allowed calculation of the kinetic parameters V_M and K_m.[4]

III. RESULTS

A. Oxygen Consumption of Control and Ecdysterone-Treated Cells

In control cells, the maximal rate of oxygen consumption (V_M) was found to be equal to 0.2 ± 0.05 fmol$\cdot$min$^{-1}\cdot$cell^{-1}, and the K_m was 4 ± 1 μM. These values show, first, that the cells are well protected against hypoxia. Indeed, to obtain a twofold decrease of the V_M, the O_2 partial pressure has to be diminished 250 $\mu M/4$ μM, or 62-fold. Second, knowing that 1 l of O_2 corresponds to the biological production of 4.7×10^3 cal, the value of the V_M allowed us to estimate the energy production as $W_o = 2 \times 10^{-11}$ cal per one cell during 1 min, or, as the synthesis of about 1.3 fmol ATP $\cdot$ min$^{-1}\cdot$ cell^{-1}.[5] All these results were found and discussed in Peronnet et al.[2]

Measuring the kinetic parameters of the O_2 consumption of cells treated with 500 nM ecdysterone for 2 or 3 d, we found that the V_M could be increased up to threefold, whereas the K_m did not vary. The energy production in the presence of the hormone is, thus, $W_H = 6 \times 10^{-11}$ cal $\cdot$ min$^{-1}\cdot$ cell^{-1}. We propose that the increase of energy $\Delta W = W_H - W_o = 4 \times 10^{-11}$ cal $\cdot$ min$^{-1}\cdot$ cell^{-1}, or 2.6 fmol ATP $\cdot$ min$^{-1}\cdot$ cell^{-1}, which is the energy required by the cells to begin their differentiation under the hormonal stimulation.

B. Effect of Different Conditions of Oxygenation on Differentiation

We verified the necessity of oxygen during differentiation by first adapting cells to grow in ordinary flasks containing different gas phases: air ($N_2 + O_2$, 150 mmHg), nitrogen ($N_2 + O_2$ traces), 50% O_2 ($N_2 + O_2$, 300 mmHg), or in "Petriperm" flasks allowing gas exchanges. Then we treated them with ecdysterone in order to check their capacity of differentiation. The criteria of differentiation retained were the induction of enzymatic activities (acetylcholinesterase, beta-galactosidase, catalase) and the increase of the rate of synthesis of actin by ecdysterone.

As expected, the enzymatic induction was poor with ecdysterone-treated cells grown with a nitrogen gas phase. On the contrary, when extra oxygenation was supplied (50% O_2 or "Petriperm" flasks), the induction was greater than in "normal" conditions (for details, see Peronnet et al.[2]).

The rate of synthesis of actin was scarcely increased by ecdysterone when the gas phase was nitrogen whereas the synthesis was greatly enhanced in good oxygenation conditions.

C. Adaptation of the Cells to Oxygen Toxicity in the Case of Ecdysterone-Induced Differentiation

We know that oxygen is both indispensable and toxic to all aerobic living things.[6] The toxicity is due to the products of O_2 reduction, such as the superoxide ion O_2^- and hydrogen peroxide H_2O_2 which are highly reactive.[7] Among the defense mechanisms of *Drosophila* cells are the enzymes superoxide dismutase and catalase. Ecdysterone was shown to induce catalase in *Drosophila*-responsive cells.[8] Also, superoxide dismutase activity was found to increase up to 2.5-fold in ecdysterone-treated cells.

Thus, the following scheme is plausible: ecdysterone induces the differentiation of *Drosophila* cells *in vitro*; this differentiation requires an increase of energy and an increase of O_2 uptake. The cells avoid the danger of the consequent overproduction of products of O_2 reduction by an increased activity of the defense enzymes superoxide dismutase and catalase.

D. Adaptation of the Cells to Oxygen Toxicity in the Case of Oxidative Stress

As a control, we wanted to see what happened when the cells were confronted with a sudden increase of O_2 uptake without enough time to increase enzymatic defenses.

Such was the case when the cells were cultured for 24 h in an anoxic state (nitrogen) and then put back in normal conditions (air). We found that in the hours following the return to normoxia: (1) the V_M doubled while the K_m remained constant, (2) the superoxide dismutase and catalase activities did not increase, (3) the synthesis of the heat shock proteins (hsp) occurred although the culture temperature remained constant (23°C). We propose[4] that the overproduction of the O_2 reduction products (superoxide ion, hydrogen peroxide) resulting from the increase of O_2 uptake might play a role in the chain of events leading to the induction of the synthesis of hsp which are supposed to protect the cells in different cases of stress.[9]

In order to test this hypothesis, we treated Drosophila cells with hydrogen peroxide (1 mM). Surprisingly, we observed an important enhancement of the synthesis of actin which was maximal 3 h after the H_2O_2 treatment, even if this treatment lasted only 10 min. The meaning of this result if not yet clear. However, as expected, several hsp were induced in the first hours following the treatment: hsp 70, 68, and 23. In fact, when H_2O_2 was used alone, the reproducibility of induction of hsp was only 50%, probably owing to the fact that the cells have a basic catalase activity which is not negligible (0.2 Sigma U per 10^6 cells). When the cells were treated with 1 mM H_2O_2 plus 1 mM aminotriazole (an inhibitor of catalase), the reproducibility was 100%. Moreover, preliminary results showed that several hsp genes were transcriptionally activated within 10 min of H_2O_2 treatment.

IV. DISCUSSION

If we compare the oxidative stress and the first steps of ecdysterone-induced differentiation in our *in vitro* Drosophila cell system, we find a common point: the increase of oxygen uptake and, consequently, the increase of the products of oxygen reduction.

In the case of ecdysterone action, this increase of toxic products is balanced by an increased induction of the enzymatic defenses. Therefore, the differentiation process can continue. In fact, the differentiation of *Drosophila* cells (morphological modifications, enzymatic and protein inductions) goes on for several days *in vitro*.

In the case of oxidative stress, the excess of oxygen uptake leads to the synthesis of hsp. This is a transitory reaction to short, abnormal situations; if the stress continues, (for instance, if the cells are cultured with pure oxygen for 24 h), the cells die.

We have seen that synthesis of hsp occurs when the natural enzymatic defense is overpassed by the excess of reactive products of O_2 reduction. We put forward the hypothesis that these products (for instance, the superoxide ion and/or hydrogen peroxide) react with a cellular compound[10] which in its turn induces the hsp.[4] Our experimental data seem in agreement with this hypothesis. According to another theory,[11] an oxidative stress induces an increase of dinucleoside polyphosphates which could act as "alarmones" in inducing synthesis of hsp. We measured the intracellular level of AP_4A, AP_4G, AP_3A, and AP_3G in control and stressed *Drosophila* cells.[12] We observed an increase of the dinucleoside polyphosphates, but the kinetics of induction of these products did not allow us to consider them as the inducers of hsp.

Another common point between the action of ecdysterone and the action of H_2O_2 was the increased synthesis of actin and of hsp 23. Ecdysterone was shown to induce the polymerization and neosynthesis of actin,[13] and this induction was correlated with the morphological processes of ecdysterone-induced cellular differentiation: elongation, increased motility, and formation of aggregates. Such morphological modifications do not occur in H_2O_2-treated cells. Increased synthesis of actin might be interpreted as a repair mechanism after the disruption of the cytoskeleton by H_2O_2.[14]

Ecdysterone is known to induce the synthesis of small molecular weight hsp,[15] which also appear in the course of development.[16] However, contrary to all stress situations, ecdysterone, to our knowledge, does not induce the synthesis of hsp 70, which is the most abundant and most conserved hsp. This essential difference might be due precisely to a better control of the enzymatic defense of the cells against the products of O_2 reduction in the case of cellular differentiation induced by a physiological hormone.

REFERENCES

1. **Best-Belpomme, M. and Courgeon, A. M.,** Présence ou absence de récepteurs saturables de l'ecdysterone dans des clones sensibles ou résistants de *Drosophila melanogaster* en culture in vitro, *C.R. Acad. Sci. Paris*, 280, 1397, 1975.
2. **Peronnet, F., Ropp, M., Rollet, E., Becker, J. L., Becker, J., Maisonhaute, C., Pernodet, J. L., Vuillaume, M., Courgeon, A. M., Echalier, G., and Best-Belpomme, M.,** Drosophila cells in culture as a model for the study of ecdysteroid action, in *Techniques in the Life Science*, Vol. C2, Kurstak, E., Ed., Elsevier Scientific Publishers, County Clare, Ireland, 1986, 1.
3. **Echalier, G. and Ohanessian, A.,** In vitro culture of Drosophila melanogaster embryonic cells, *In Vitro*, 6, 162, 1970.
4. **Ropp, M., Courgeon, A. M., Calvayrac, R., and Best-Belpomme, M.,** The possible role of the superoxide ion in the induction of heat shock and specific proteins in aerobic *Drosophila* cells during return to normoxia after a period of anaerobiosis, *Can. J. Biochem. Cell. Biol.*, 61, 456, 1983.
5. **Jöbsis, F. F.,** Basic processes in cellular respiration, in *Handbook of Physiology*, Vol. 1 (Sect. 3), Fenn, W. O., and Rahn, H., Eds., American Physiology Society, Washington, D.C., 1964, 63.
6. **Fridovich, I.,** Oxygen radicals, hydrogen peroxide and oxygen toxicity, in *Free Radicals in Biology*, Vol. 1, Pryor, W. A., Ed., Academic Press, New York, 1976, 238.
7. **Fridovich, I.,** Superoxide radical and superoxide dismutases, in *Oxygen and Living Processes: An Interdisciplinary Approach*, Gilbert, D. L., Ed., Springer-Verlag, New York, 1981, 250.
8. **Best-Belpomme, M. and Ropp, M.,** Catalase is induced by ecdysterone and ethanol in *Drosophila* cells, *Eur. J. Biochem.*, 121, 349, 1982.
9. **Schlesinger, M. J., Ashburner, M., and Tissières, A.,** *Heat Shock: From Bacteria to Man*, Cold Spring Harbor Laboratory Press, Cold Spring Harbor, NY, 1982.

10. **Compton, J. L. and McCarthy, B. J.,** Induction of the *Drosophila* heat shock response in isolated polytene nuclei, *Cell,* 14, 191, 1978.
11. **Lee, P. C., Bochner, B. R., and Ames, B. N.,** AppppA, heat shock stress, and cell oxidation, *Proc. Natl. Acad. Sci. U.S.A.,* 90, 7496, 1983.
12. **Brevet, A., Plateau, P., Best-Belpomme, M., and Blanquet, S.,** Variation of Ap4A and other dinucleoside polyphosphates in stressed *Drosophila* cells, *J. Biol. Chem.,* 260, 15566, 1985.
13. **Couderc, J. L., Cadic, A. L., Sobrier, M. L., and Dastugue, B.,** Ecdysterone induction of actin synthesis and polymerization in a *Drosophila melanogaster* cultured cell line, *Biochem. Biophys. Res. Commun.,* 102, 188, 1982.
14. **Schraufstatter, M. J., Hyslop, P. A., Hinshaw, D. B., Spragg, R. G., Sklar, L. A., and Cochrane, C. G.,** Hydrogen peroxide-induced injury of cells and its prevention by inhibitors of poly(ADP-ribose) polymerase, *Proc. Natl. Acad. Sci. U.S.A.,* 83, 4908, 1986.
15. **Ireland, R. C. and Berger, E. M.,** Synthesis of low molecular weight heat shock peptides stimulated by ecdysterone in a cultured *Drosophila* cell line, *Proc. Natl. Acad. Sci. U.S.A.,* 79, 855, 1982.
16. **Cheney, C. M. and Shearn, A.,** Developmental regulation of *Drosophila* imaginal disc proteins: synthesis of a heat shock protein under non-heat-shock conditions, *Dev. Biol.,* 95, 325, 1983.

Chapter 10

METABOLISM OF ECDYSTEROIDS BY INSECT TISSUES *IN VITRO*

Catherine Blais, Jean-François Modde, Philippe Beydon, and René Lafont

TABLE OF CONTENTS

I. INTRODUCTION

Ecdysone, which is secreted by prothoracic glands into hemolymph, is converted first into 20-hydroxyecdysone (generally considered as the active hormone), and then into a lot of various metabolites which are much less active and, thus, might represent inactivation products (see reviews by Lafont and Koolman[1] and Koolman and Karlson[2]). The reactions involve several positions of the ecdysone molecule which are often regarded as essential for hormonal activity. These are the secondary alcohol functions at C-2, C-3, and C-22, and the methyl group at C-26 which is first converted into a primary alcohol, then into a carboxylic acid.[3]

The present paper is especially concerned with the tissue localization of these conversions and summarizes results obtained in the authors' laboratory on two insect species, i.e., one heterometabolous species, *Locusta migratoria*, and one holometabolous species, *Pieris brassicae*.

II. MATERIALS AND METHODS

Experimental procedures have been described previously.[3-9] Tissues were dissected into an insect saline and thereafter incubated with [³H]ecdysone in culture medium:medium S29 (gift from Dr. J.-C. Landureau) for *Locusta* and Grace's medium (Gibco) for *Pieris* tissues. After different incubation times (1, 6, and 24 h), tissues and medium were processed together and analyzed by high-performance liquid chromatography (HPLC) in various systems. Some metabolites were identified by coelution with reference compounds in at least two chromatographic systems. New compounds were isolated and identified by a combination of enzymatic hydrolyses (conjugates), double labeling (phosphates), mass spectrometry (CI/ D,FAB), proton-nuclear magnetic resonance (NMR), and various derivative procedures (e.g., formation of methyl esters, acetates, and acetonides).

III. RESULTS AND DISCUSSION

The different metabolic reactions affecting the ecdysone molecule and detected after incubation of tissues are represented in Figure 1.

The results obtained from incubation of *Locusta* tissues in the presence of [³H]ecdysone are summarized in Table 1 (from Modde et al.,[5]). Ecdysone 20-hydroxylation occurred in all the tissues assayed. Malpighian tubules and, to a lesser extent, gut were active organs at all stages tested. Acetylation at C-3 appeared to occur mainly in the guts of larvae; in adults, however, fat body and epidermis became able to perform acetylation. Phosphate conjugates were formed both by larval and adult tissues, adult Malpighian tubules being very active. Ecdysonoic acids were detected after incubations of all tissues, but always at a low level (results not shown). This reaction did not appear to be characteristic of a tissue or a developmental stage.

Incubation of *Pieris* tissues with [³H]ecdysone also led to the formation of various metabolites, their nature and proportion depending both on the tissue and on the animal stage (Table 2).

Formation of 20-hydroxyecdysone was detected in all tissues, mainly in fat body and imaginal wing disks. The major site of the reaction was fat body in prepupae and imaginal disks in pupae. These results are in agreement with those obtained from incubations of subcellular fractions.[8] 3-Epimers are only formed by gut, at all stages assayed, as was shown from incubations of subcellular fractions.[7,10] 3-Epimerization is predominant in larvae. In prepupae and pupae, the proportion of 3-epimer forms is reduced and 3-oxo-compounds accumulate; these compounds are intermediates between 3β- and 3α-hydroxy-ecdysteroids.[7]

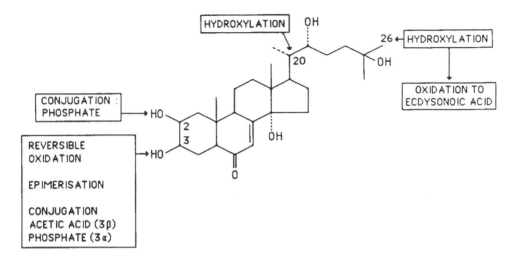

FIGURE 1. Summary of the metabolic reactions affecting the ecdysone molecule in *Locusta migratoria* and *Pieris brassicae*.

They are also formed by fat body. The last metabolic transformation of ecdysone detected *in vitro* is 26-hydroxylation. It is only performed by imaginal disks. 26-Hydroxyecdysone is a precursor of ecdysonoic acid[3] (see Figure 1), which was not detected in the present experiments. It is formed, however, by imaginal disks after long-term incubations.[3,6]

Ecdysone 20-hydroxylation is a general reaction *in vitro* as well as *in vivo*. This reaction has been characterized *in vitro* in several insect species and different tissues (see references in Lafont and Koolman,[1] and Smith,[11]). It is mainly localized in Malpighian tubules in *Locusta*,[12] in fat body or wing disks in *Pieris*. The major difference between the two species studied concerns ecdysone conjugates, not detected in *Pieris* tissue incubations, although they are formed *in vivo* after injection of tritiated ecdysone into larvae, prepupae, or pupae.[9,13] For *Locusta*, ecdysteroid conjugates could be formed *in vitro* as well as *in vivo*.[5,12] Thus, some differences may appear concerning the metabolic patterns obtained with the two experimental approaches. For *Pieris* and *Locusta*, ecdysonoic acids are always formed *in vivo*, at any stage. On the other hand, 3-dehydro compounds are not detected in *Pieris* prepupae nor in young pupae injected with labeled ecdysteroids.[9,13] As the formation of 3-dehydro compounds from 3β-hydroxyecdysteroids is reversible,[7] it is possible that *in vivo* conditions favor the reduced forms, whereas the present *in vitro* conditions allow 3-dehydro compounds to accumulate. Thus, our *in vitro* conditions (for *Pieris*) do not allow all the metabolic reactions of ecdysteroids to take place at a "normal" rate, and this could be due to culture medium inadequacy. On the other hand, *in vitro* experiments are of interest to determine the tissue distribution of some reactions and their evolution throughout insect development. Our results show that the major site of a given reaction may vary during development. However, nothing is known at the present time regarding the mechanisms which could control the tissue specificity and the variations of ecdysone metabolism.

Table 1
METABOLISM OF ECDYSONE BY *LOCUSTA MIGRATORIA* TISSUES *IN VITRO*

Age of donor	Tissue	Formation of metabolites			
		20-Hydroxyecdysone	3-Acetates	Phosphate conjugates	
Fifth larval instar (day 2)	Malpighian tubules	+ +	N.D.	+	
	Gut	+	+	+	
	Fat body	+	N.D.	+	
	Epidermis	(+)	N.D.	+	
Fifth larval instar (day 7)	Malpighian tubules	+ +	N.D.	N.D.	
	Gut	+	+ +	+ +	
	Fat body	(+)	N.D.	N.D.	
	Epidermis	(+)	N.D.	N.D.	
Adult (males)	Malpighian tubules	+	N.D.	+ +	
	Gut	(+)	+ +	+ +	
	Fat body	N.D.	+ +	+ +	
	Epidermis	(+)	+	+	

Note: Incubation took place in the presence of [^{3}H]ecdysone (2×10^{-7} M) in medium S 29; N.D. = not detected; % conversion is indicated as follows: (+) = <1; + = >1 but <10; + + = >10.

From Modde, J.-F., Lafont, R., and Hoffman, J. A., *Int. J. Invertebr. Reprod. Dev.*, 7, 161, 1984. With permission.

Table 2
METABOLISM OF ECDYSONE BY *PIERIS BRASSICAE* TISSUES *IN VITRO*

Age of donor	Tissue	Formation of metabolites			
		20E[a]	3-Epimers	3-Dehydro compounds	26E[b]
Fifth larval instar (late)	Imaginal wing disks	+	N.D.	N.D.	N.D.
	Fat body	+	N.D.	N.D.	N.D.
	Gut (+ Malpighian tubules)	(+)	++	N.D.	N.D.
	Epidermis	(+)	N.D.	N.D.	N.D.
Prepupa	Imaginal wing disks	(+)	N.D.	N.D.	+
	Fat body	++	N.D.	N.D.	N.D.
	Gut (+ Malpighian tubules)	+	+	++	N.D.
	Epidermis	(+)	N.D.	N.D.	N.D.
Pupa (day 4)	Wing disks	++	N.D.	N.D.	N.D.
	Fat body	(+)	N.D.	+	N.D.
	Gut	(+)	(+)	+	N.D.
Pupa (day 9)	Wing disks	+	N.D.	N.D.	+
	Fat body	+	N.D.	++	N.D.
	Gut	N.D.	+	++	N.D.

Note: Incubation took place in the presence of [^{3}H]ecdysone (2×10^{-7} M) in Grace's medium; N.D. = not detected; % conversion is indicated as follows: (+) = <1; + = >1 but <10; ++ = >10.

[a] 20E = 20-hydroxyecdysone.
[b] 26E = 26-hydroxyecdysone.

REFERENCES

1. **Lafont, R. and Koolman, J.,** Ecdysone metabolism, in *Biosynthesis, Metabolism and Mode of Action of Invertebrate Hormones,* Hoffmann, J. and Porchet, M., Eds., Springer-Verlag, Berlin, 1984, 196.

2. **Koolman, J. and Karlson, P.,** Regulation of ecdysteroid titer: degradation, in *Comprehensive Insect Physiology, Biochemistry and Pharmacology,* Vol. 7, Kerkut, G. A. and Gilbert, L. I., Eds., Pergamon Press, Oxford, 1985, 343.

3. **Lafont, R., Blais, C., Beydon, P., Modde, J.-F., Enderle, U., and Koolman, J.,** Conversion of ecdysone and 20-hydroxyecdysone into 26-oic derivatives is a major pathway in larvae and pupae of species from three insect orders, *Arch. Insect Biochem. Physiol.,* 1, 41, 1983.

4. **Lafont, R., Pennetier, J.-L., Andrianjafintrimo, M., Claret, J., Modde, J.-F., and Blais, C.,** Sample processing for high-performance liquid chromatography of ecdysteroids, *J. Chromatogr.,* 236, 137, 1982.

5. **Modde, J.-F., Lafont, R., and Hoffmann, J. A.,** Ecdysone metabolism in *Locusta migratoria* larvae and adults, *Int. J. Invertebr. Reprod. Dev.,* 7, 161, 1984.

6. **Blais, C. and Lafont, R.,** *In vitro* differentiation of *Pieris brassicae* imaginal discs: effects and metabolism of ecdysone and ecdysterone, *Wilhelm Roux Arch. Entwicklungsmech. Org.,* 188, 27, 1980.

7. **Blais, C. and Lafont, R.,** Ecdysteroid metabolism by soluble enzymes from an insect. Metabolic relationships between 3β, 3α-hydroxy- and 3-oxoecdysteroids, *Hoppe Seylers Z. Physiol. Chem.,* 365, 809, 1984.

8. **Blais, C. and Lafont, R.,** Ecdysone-20-hydroxylation in imaginal wing discs of *Pieris brassicae* (Lepidoptera): correlations with ecdysone and 20-hydroxyecdysone titers in pupae, *Arch. Insect Biochem. Physiol.,* 3, 501, 1986.

9. **Beydon, P., Girault, J.-P., and Lafont, R.,** Ecdysone metabolism in *Pieris brassicae* during the feeding last larval instar, *Arch. Insect Biochem. Physiol.,* 4, 139, 1987.

10. **Nigg, H. N., Svoboda, J. A., Thompson, M. J., Kaplanis, J. N., Dutky, S. R., and Robbins, W. E.,** Ecdysone 20-hydroxylation from the midgut of the tobacco hornworm (*Manduca sexta* L.), *Lipids,* 9, 971, 1974.

11. **Smith, S. L.,** Regulation of ecdysteroid titer: synthesis, in *Comprehensive Insect Physiology, Biochemistry and Pharmacology,* Vol. 7, Kerkut, G. A. and Gilbert, L. I., Eds., Pergamon Press, Oxford, 1985, 295.

12. **Feyereisen, R., Lagueux, M., and Hoffmann, J. A.,** Dynamics of ecdysone metabolism after ingestion and injection in *Locusta migratoria, Gen. Comp. Endocrinol.,* 29, 319, 1976.

13. **Beydon, P., Claret, J., Porcheron, P., and Lafont, R.,** Biosynthesis and inactivation of ecdysone during the pupal adult development of the cabbage butterfly, *Pieris brassicae, Steroids,* 38, 633, 1981.

Chapter 11

TESTES OF THE TOBACCO BUDWORM MOTH: ECDYSTEROID PRODUCTION BY THE TESTIS SHEATH

M. J. Loeb, E. P. Brandt, and C. W. Woods

TABLE OF CONTENTS

I. INTRODUCTION

The testes of Lepidoptera are usually ovoid, paired organs suspended in the abdominal cavity. The sheath surrounding each testis has two prominent cellular layers. The outer cells cover the circumference of the testis, while the inner layer also extends into the lumen to form the walls of the four follicles. Germ cells develop as clones within each follicle lumen.[1] Shortly before pupation, the testes of many species of Lepidoptera fuse and twist to form a single organ. This process requires ecdysteroid.[2] Male tract development is also dependent on the presence of ecdysteroid.[3] Release of sperm from the testis to the male genital tract in adult male Lepidoptera is negatively controlled by ecdysteroid.[4]

II. TESTES OF *HELIOTHIS VIRESCENS* AS A SOURCE OF ECDYSTEROIDS

Ecdysteroid radioimmunoassay (RIA)[5] of testes and of the media in which testes were cultured (using ecdysteroid antibody provided by W. E. Bollenbacher) is shown in Figure 1. The difference between the curves indicates spontaneous immunodetectable ecdysteroid synthesis late in the last larval instar, in midpupal development, and a small amount in the new adult.[6]

Testes of late-last instar larvae were separated into components by gentle teasing and agitation in Ringer solution. The components were then washed. RIA indicated approximately 40% of immunoreactive ecdysteroid in the sheaths, 53% in the follicle fluid, and 7% in the cysts.[7] Thus, immunoreactive ecdysteroids are sequestered as well as released by testes. Only whole testes and isolated testis sheaths were capable of ecdysteroid synthesis *in vitro*,[7] indicating testis sheaths as the site of synthesis. Immunocytology of testes with antiecdysteroid antibody (stained with horseradish peroxidase and diaminobenzoic acid) supported this view. Testes from late-last instar larvae, rich in RIA-detectable ecdysteroid (Figure 1), showed dense immunostaining in and around cells of the inner testis sheath, as shown in Figure 2.[8] Frequent washing during the procedure may have removed unbound immunoreactive ecdysteroid expected in the follicle lumen.

Brain peptides induce ecdysteroid secretion by prothoracic glands (prothoracicotropic hormone [PTTH]),[9,10] and ovaries (egg development neurohormone [EDNH]).[11,12] Methanol extracts of brains of late-last instar and midpupal male *H. virescens*, incubated with testes of early-last instar larvae, initiated ecdysteroid production in these normally inactive organs.[13] The developmental periods when brains were capable of inducing ecdysteroid synthesis coincided with spontaneous testis secretory activity. The testis ecdysiotropin (TE) activity was dose dependent (Figure 3). Physical properties of TE set it apart from PTTH[10] and EDNH.[12] TE was labile when heated, frozen, or stored dry or in aqueous media, although it kept well in acidic methanol at $-20°C$. Like PTTH and EDNH, TE activity was destroyed by proteases; it is probably a peptide. In size exclusion chromatography (Waters I-125 column), TE comigrated with markers cytochrome *c* and insulin, suggesting a molecular weight between 6 and 12 kDa.[13] TE may be a discrete ecdysiotropin acting on testes of male Lepidoptera.

Isolated sheaths from late-last instar testes, which normally produce ecdysteroid spontaneously *in vitro*, showed a startling dose-dependent positive response to exogenous 20-hydroxyecdysone (Figure 4).[14] Young testes, which do not normally produce ecdysteroid spontaneously *in vitro*, were not affected by exogenous ecdysteroids, although they could be stimulated to produce immunodetectable ecdysteroids after exposure to TE. The data suggest that once the mechanism for ecdysteroid production is activated by TE, the amount of ecdysteroid actually produced can be modulated by exogenous ecdysteroid.

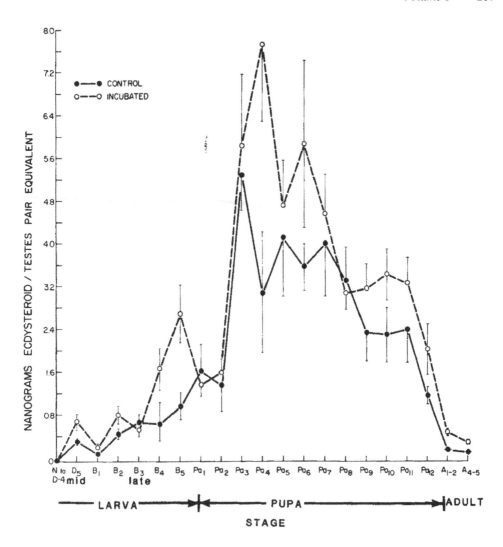

FIGURE 1. Changes in immunodetectable ecdysteroid content of testes of *H. virescens* from mid-last larval instar to adult. Control (———) compared to 2.5-h-incubated testes plus medium (---).

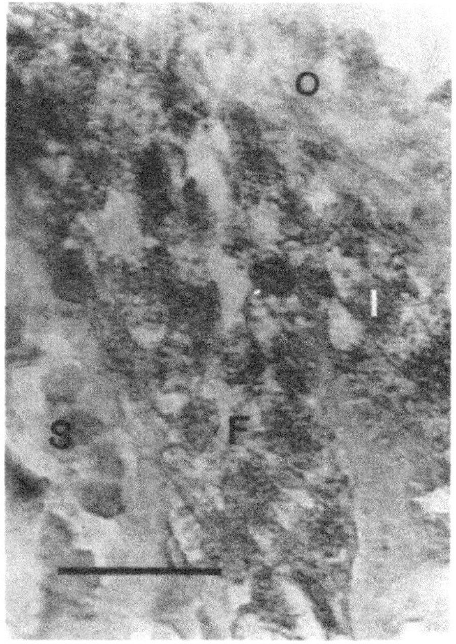

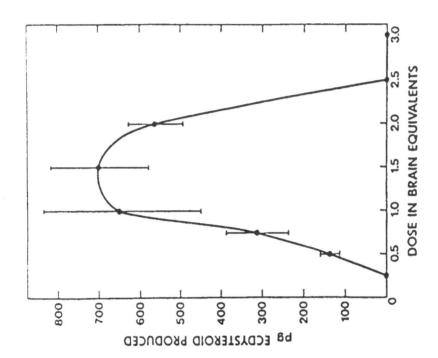

FIGURE 3.   Dose response to testis ecdysiotropin in brain extract. Assay performed on testes from early mid-last instar of *H. virescens* larvae. Extract derived from brains of mid-development pupae.

FIGURE 2.   Section of testis from a late-last instar larva incubated with antiecdysteroid and DAB. F indicates follicle wall; I, inner testis sheath; L, follicle lumen; O, outer sheath; S, spermatocyst. Bar = 20 μm.

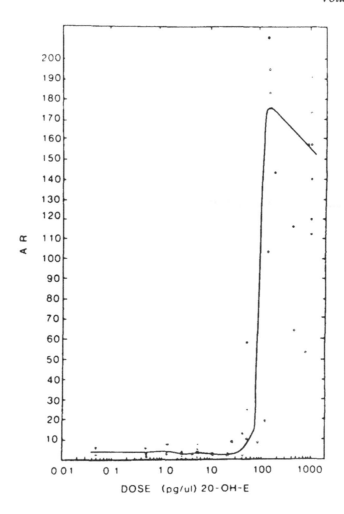

FIGURE 4. Activation of immunodetectable endogenous ecdysteroid production by increasing titers of exogenous 20-hydroxyecdysone incubated with testis sheaths of late-last instar *H. virescens*. AR = net immunoreactive ecdysteroid detected after incubation in 20-hydroxyecdysone/immunoreactive ecdysteroid produced in Ringer.

REFERENCES

1. **Szollozi, A.,** Relationships between germ and somatic cells in the testes of locusts and moths, in *Insect Ultrastructure,* King, R. C. and Akai, H., Eds., Plenum Press, New York, 1982, 32.
2. **Nowock, J.,** Induction of imaginal differentiation by ecdysone in testes of *Ephestia kuhniella, J. Insect Physiol.,* 18, 1699, 1972.
3. **Szopa, T. M., Rousseaux, J.-J., Yuncker, C., and Happ, G. M.,** Ecdysteroids accelerate mitoses in accessory glands of beetle pupae, *Dev. Biol.,* 107, 325, 1985.
4. **Thorson, B. J. and Riemann, J. G.,** Effects of 20-hydroxyecdysone on sperm release from the testes of the Mediterranean flour moth, *Anagasta kuehniella* (Zeller), *J. Insect Physiol.,* 28, 1013, 1982.
5. **Borst, D. W. and O'Connor, J. D.,** Trace analysis of ecdysones by gas-liquid chromatography, radioimmunoassay and bioassay, *Steroids,* 24, 637, 1974.
6. **Loeb, M. J., Brandt, E. P., and Birnbaum, M. J.,** Ecdysteroid production by testes of the tobacco budworm, *Heliothis virescens* from last larval instar to adult, *J. Insect Physiol.,* 30, 375, 1984.

7. **Loeb, M. J., Woods, C. W., Brandt, E. P., and Borkovec, A. B.,** Larval tests of the tobacco budworm: a new source of insect ecdysteroids, *Science,* 218, 896, 1982.

8. **Loeb, M. J.,** Ecdysteroids in testis sheaths of *Heliothis virescens* larvae: an immunocytochemical study, *Arch. Insect Biochem. Physiol.,* 3, 173, 1986.

9. **Williams, C. M.,** Physiology of insect diapause. II. Interaction between the pupal brain and prothoracic glands in the metamorphosis of the great silkworm, *Platysamia cecropia, Biol. Bull.,* 93, 89, 1947.

10. **Bollenbacher, W. E., Agui, N., Granger, N. A., and Gilbert, L. I.,** Insect prothoracic glands *in vitro*: a system for studying the prothoracicotropic hormone, in *Invertebrate Systems in Vitro,* Kurstak, E., Maramorosch, K. K., and Dübendorfer, A., Eds., Elsevier/North-Holland, Amsterdam, 1980, 253.

11. **Lea, A.,** Regulation of egg maturation in the mosquito by the neurosecretory system: the role of the corpus cardiacum, *Gen. Comp. Endocrinol. Suppl.,* 3, 602, 1972.

12. **Hagedorn, H. H., Shapiro, J. P., and Hanaoka, K.,** Ovarian ecdysone secretion is controlled by a brain hormone in an adult mosquito, *Nature (London),* 282, 92, 1979.

13. **Loeb, M. J., Woods, C. W., Brandt, E. P., and Borkovec, A. B.,** Factor from brains of male *Heliothis virescens* induces ecdysteroid production by testes, in *Neurochemistry and Neurophysiology 1986,* Borkovec, A. B. and Gelman, D. B., Eds., Humana Press, Clifton, NJ, 1986, 367.

14. **Loeb, M. J., Brandt, E. P., and Woods, C. W.,** Effect of exogenous ecdysteroid titer on endogenous ecdysteroid production *in vitro* by testes of the tobacco budworm, *Heliothis virescens, Zoology,* 240, 75, 1986.

Chapter 12

IN VITRO RELEASE OF PROTHORACICOTROPIC HORMONE (PTTH) FROM THE BRAIN OF *MAMESTRA BRASSICAE* L.: EFFECTS OF NEUROTRANSMITTER SUBSTANCES ON PTTH RELEASE

Noriaki Agui

TABLE OF CONTENTS

I. INTRODUCTION

The morphological analogy between the hypothalamo-pituitary complex of vertebrates and the pars intercerebralis-retrocerebral complex of insects has been shown in early studies on neurosecretion.[1,2] Namely, both systems are composed of several groups of neurosecretory cells located in specific areas of the brain, and these neurosecretory cells are linked to neurohemal organs containing both extrinsic neurosecretory axons and intrinsic cells. In the vertebrate nervous system, it is well established that biogenic amines regulate the release of peptidergic hormones from the hypothalamo-pituitary complex.[3,4] In insects many transmitter substances such as monoamines, acetylcholine, and gamma-aminobutyric acid (GABA) are known as putative neurotransmitters of nervous systems.[5,6] However, in comparison to the plentiful information on the hypothalamo-pituitary complex of vertebrates, relatively little is known of the following: (1) the functional significance of the central aminergic, cholinergic, and amino acidergic system and (2) the control of the level of these neurotransmitter substances in pars intercerebralis-retrocerebral complexes in insects.

Organ culture methodology for the insect central nervous system, such as the pars intercerebralis-retrocerebral complex (corpus cardiacum-corpus allatum complex) provides more precise information about intrinsic and extrinsic regulation of specific neurosecretory cells and their neurohemal organs. In early experiments, the secretory activity of prothoracicotropic hormone (PTTH) in the brain had been investigated by coculturing brains with the PTTH target organ, the prothoracic glands (PG).[7,8] The improved *in vitro* PG assay for PTTH[9] provided further evidence that PTTH activity, either in brain alone or in brain-retrocerebral complexes, could be measured more accurately under specific culture conditions.[10-15]

In the present studies, an attempt was made to investigate the role of neurotransmitter substances in the regulation of peptidergic neurosecretory cell (PTTH cell) activities by the culture of isolated brain of the cabbage armyworm, *Mamestra brassicae*.

II. MATERIALS AND METHODS

A. Experimental Animals

Larvae of the cabbage armyworm, *Mamestra brassicae* L. (Lepidoptera: Noctuidae), were reared on an artificial diet[16] at 25°C under a long-day photoperiod (LD, 16:8). Fresh day-0 pupae within 0 to 4 h after pupation were used as donors for the culture of brains, since the critical period of pupal PTTH release for adult development was at 3 d after pupation.[17] The prothoracic glands used in the *in vitro* PTTH assay were also taken from fresh pupae, thus facilitating accurate stage selection for the PTTH assay.[9,10]

B. Chemicals

All transmitter substances, their agonists, and antagonists used in the experiment were purchased from Sigma Co., U.S. The group of catecholamines — dopamine (DA), norepinephrine (NEP), and epinephrine (EP) — and the group of indole amines such as serotonin (5-HT) were used as biogenic monoamines. Acetylcholine (ACh), γ-aminobutyric acid (GABA), and c-AMP were also used in the experiment as typical neurotransmitters and second messengers of the insect nervous system. Nialamide, which acts as an inhibitor of monoamine oxidase, and eserine and diisopropylphosphorofluoridate (DFP) which act as reversible or irreversible blockers of acetylcholinesterase, were used as agonists of the neurotransmitters. Furthermore, hemicolinium-3, which acts to block the uptake of choline, was used as an antagonist of ACh. All chemicals were dissolved in Grace's medium (Grand Island Biological Co., U.S.) at appropriate concentrations (5 to 500 μM) to investigate their effect either on ecdysone synthesis of the cultured PG or on PTTH release from the cultured brain *in vitro*.

C. Culture Methods

On day 0, *Mamestra* pupae were submerged in 70% ethanol for 5 s and then sterilized in 0.1% $HgCl_2$ for 3 min. Pupae were then rinsed three times with sterilized water, and their brains were dissected out and washed twice with Grace's medium. Four brains were cultured together in a standing drop (50 μl) of Grace's medium (in the presence or absence of appropriate concentrations of neurotransmitters, their agonists, or their antagonists) in a four-well tissue culture plate (well diameter: 15 mm; Nunc, Denmark) for 24 h. The spaces between the four wells in the plate were filled with 3 ml of sterilized water to prevent evaporation of culture medium, and cultures were kept at 25°C under high humidity, continuous light, and an air atmosphere.

D. Assay of PTTH Activity

After culturing, brains were homogenized in Grace's medium (one brain per 50 μl) in a microglass homogenizer (0.5 ml) with a Teflon® pestle (Bellco Glass Inc., U.S.), heat treated at 100°C for 2 min followed by centrifugation at 8000 × *g* for 5 min. The PTTH activity in the resultant supernatant was assayed by the *in vitro* PG assay method. The PTTH activities of aliquots from incubation media were also determined using the *in vitro* PG assay. The *in vitro* PTTH assay by *Mamestra* PG[10] was performed using the method previously described for *Manduca*,[9] with minor modifications. The PG of fresh pupae were extirpated in 0.9% NaCl solution and then rinsed for 20 to 30 min in Grace's medium to remove hemolymph ecdysteroids. After washing, the individual glands were transferred either to standing drops (25 μl) of Grace's medium containing appropriate chemicals for controls or to drops of the same volume of harvested or tissue-homogenized medium for tests. All were placed in plastic multiwell tissue culture plates (Falcon 24-well plates purchased from Becton Dickinson, U.S.). The incubation of PG was carried out for 2 h under the same conditions as for brain cultures. The quantity of ecdysone from each culture (test PG or the contralateral control PG) was determined by ecdysone radioimmunoassay (RIA). The PTTH activity was expressed as a PG-activation ratio (Ar), which represents the quantity of ecdysone synthesized by the experimental gland during the 2-h incubation period divided by the amount synthesized by the contralateral control gland during the same period. The mean ± standard error of the activation ratios from three assays was obtained.

E. Radioimmunoassay

To quantify ecdysone synthesis by PG, an ecdysone RIA[18,19] was used. A 22-hemisuccinate derivative of ecdysone was used as the antigen to generate the antisera. This antiserum exhibits equivalency of binding for ecdysone and 20-hydroxyecdysone.[20] The labeled ligand was [23,24,^{3}H]- ecdysone at 80 Ci/mmol (New England Nuclear) and the competing unlabeled ligand was also ecdysone (Eco Chemical Intermediate).

III. RESULTS

A. Effects of Neurotransmitter Substances on the Prothoracic Gland *in Vitro*

If we want to determine the PTTH activity in brain-conditioned culture medium containing transmitter substances, their agonists, and antagonists using the *in vitro* PG assay, we should evaluate the effect of these chemicals on ecdysone synthesis in the *in vitro* PG assay. If these chemicals caused any inhibitory or stimulatory effects on the basal ecdysone synthesis of PG *in vitro*, then the *in vitro* PG assay could not yield an accurate measure of the PTTH activity in brain-conditioned culture medium containing these substances. Table 1 shows the effects of different concentrations (5, 50, and 500 μM) of these chemicals on basal ecdysone synthesis of day 0 *Mamestra* pupal PG *in vitro*. There appeared to be some inhibitory effect (Ar <1.0) on basal ecdysone synthesis of the cultured PG caused by treatment of higher

Table 1
EFFECTS OF TRANSMITTER SUBSTANCES, THEIR AGONISTS, AND ANTAGONISTS ON ECDYSONE SYNTHESIS OF PROTHORACIC GLAND *IN VITRO*

Chemicals	Activation ratio (Ar)		
	5 μM	50 μM	500 μM
Dopamine	1.0 ± 0.1	1.0 ± 0.3	0.8 ± 0.2
Norepinephrine	0.7 ± 0.3	0.7 ± 0.3	0.8 ± 0.8
Epinephrine	0.9 ± 0.1	1.1 ± 0.3	1.1 ± 0.6
Serotonin	1.0 ± 0.0	0.8 ± 0.2	0.6 ± 0.3
Acetylcholine	1.4 ± 0.4	1.0 ± 0.4	1.4 ± 0.7
GABA	0.8 ± 0.2	0.9 ± 0.3	0.9 ± 0.4
cAMP	1.2 ± 0.5	0.9 ± 0.4	0.8 ± 0.3
Nialamide	0.7 ± 0.0	0.4 ± 0.0	0.5 ± 0.0
Eserine	1.3 ± 0.3	1.0 ± 0.5	0.7 ± 0.2
DFP	0.9 ± 0.2	0.9 ± 0.1	0.9 ± 0.3
Hemicolinium-3	0.5 ± 0.2	1.0 ± 0.5	0.8 ± 0.2

Note: Activation ratio (Ar) of PTTH represented by means ± SE of the triplicated PG assay. Ar ≐ 1 means no inhibitory nor stimulatory effect of chemicals on the basal ecdysone synthesis of PG *in vitro*. $0 < Ar < 1$ means an inhibitory effect of chemicals on the basal ecdysone synthesis of PG *in vitro*.

concentrations of serotonin and nialamide. However, the other chemicals did not show any effect on ecdysone synthesis of day-0 pupal PG *in vitro*, even when the high concentration (500 μM) was used. The above results suggest that the *in vitro* PG assay method for PTTH is basically applicable for measuring the PTTH activity of brain-conditioned culture medium treated with neurotransmitters, their agonists, and antagonists. Furthermore, based on the results, the highest concentration (500 μM) of these chemicals was used in further experiments. In the assay of PTTH activity in the harvested medium using the *in vitro* PG assay, the contralateral control PG were also cultured in medium containing the same concentration of chemicals as that of test cultures.

B. Spontaneous PTTH Release from the Cultured Brain

As shown in the dose response curve of Figure 1A, there are no significant differences between PTTH activities in the extract from intact day-0 pupal *Mamestra* brains and in homogenates of cultured day-0 pupal brain. Strong PTTH activity was detected in medium which was conditioned with four brains in 50 µl Grace's medium for 24 h, when PTTH activity was measured without any dilution (Figure 1B). This activity in the harvested medium was approximately equal to the activity in the extract from an intact brain at a concentration of 0.5 brains per 50 µl of medium. This result led us to calculate that one brain can release 0.125 (0.5 brain equivalents per four brains) brain equivalents of PTTH per 24 h *in vitro*. This quantity also represents the PTTH newly synthesized by brain cultured for 24 h.

C. Effect of Transmitter Substances on PTTH Release

The dose response curve of Figure 2 shows the effect of acetylcholine (ACh) on release and synthesis of PTTH of day 0 *Mamestra* pupal brains *in vitro*. There was no significant difference between the PTTH activity of the extract from brains cultured in medium with

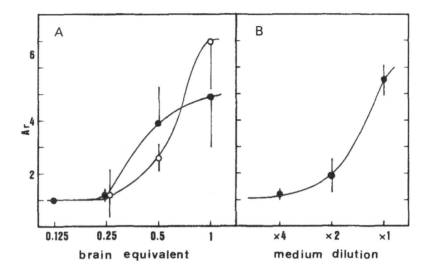

FIGURE 1. Comparison of PTTH activities in intact brains, cultured brains, and harvested medium conditioned with pupal brains. Four separate cultures composed of cocultures of four day-0 pupal brains were each incubated in 50 μl of Grace's medium for 24 h. All brains collected from separate cultures were homogenized for extraction of PTTH, and then the extract was serially diluted two times from 1 to 0.125 brain equivalents per 50 μl of medium for assaying PTTH activity. The closed circles in (A) represent the PTTH activity of cultured brains, and the open circles represent PTTH activity extracted from *in situ* day-0 pupal brains. The harvested media of four separate experiments were mixed and diluted serially 2 to 4× for assaying PTTH as shown in (B). Activation ratio (Ar) of PTTH represents mean ± SE of triplicate assays.

ACh (500 μ*M*) and in that from brain cultured without ACh (Figure 2A). In the comparative experiments of PTTH activities of brain-cultured media, there appeared to be significant differences between PTTH activities in media with ACh and those without it. Namely, an eightfold dilution of ACh-treated medium still showed higher PTTH activity than a twofold dilution of the brain-incubated medium without ACh (Figure 3B). These results indicate that ACh strongly stimulates PTTH release from cultured day-0 *Mamestra* pupal brain *in vitro*.

Using a twofold serial dilution of either the extract of the cultured brain or the harvested medium, the effects of ACh agonists, antagonists, and other transmitter substances on PTTH activity of cultured brains were further investigated as summarized in Figure 3. In the presence of monoamines (NEP and EP), monoamine agonists (NI), indole amines (5-HT), GABA, and cAMP, PTTH activity was not detected in harvested media even after diluting the media twofold; however, treatment with dopamine showed a slight increase of PTTH release into the culture medium (Figure 3A). These results demonstrate that the PTTH released from the brain in the presence of these substances was as active as the control medium conditioned with brain. On the contrary, as described in Figure 2B, very strong PTTH activity was still detected in the harvested medium after an eightfold dilution in the case of ACh treatment. In the presence of eserine and DFP which act as agonists of ACh, the cultured brain released higher PTTH activity than the brain cultured in medium alone. However, no inhibitory effect on PTTH release was observed in the treatment of hericolinium-3, which acts to block choline uptake. The results of PTTH assay against the brain-cultured media suggest that the PTTH released from cultured brains may be a mediated cholinergic neural system. The PTTH activities in brain extracts were further investigated after 24-h treatment with various transmitter substances (Figure 3B). There was no significant difference in PTTH activity between extracts of tested brains cultured with most chemicals and those of control brains cultured in medium alone. However, brains treated with cAMP showed higher PTTH activity than controls.

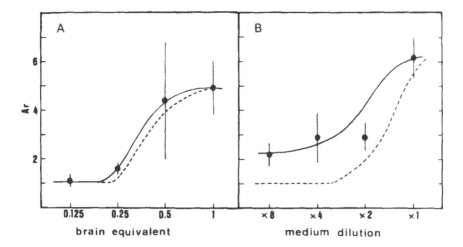

FIGURE 2.   The effect of acetylcholine (ACh) on the *in vitro* release and accumulation of PTTH in pupal brain. Four separate cultures composed of cocultures of four day-0 pupal brain were each incubated in 50 μl of medium containing 500 μ*M* ACh for 24 h, and then all collected brains and harvested media from the separate cultures were assayed for their PTTH activities by the same method as described in Figure 1. PTTH activity in brain treated with ACh is shown in (A). PTTH activity in the harvested medium conditioned with brain in the presence of ACh is shown in (B). Dotted lines in both (A) and (B) represent control-PTTH activity in the cultured brain and in the harvested medium without ACh as shown in Figure 1.

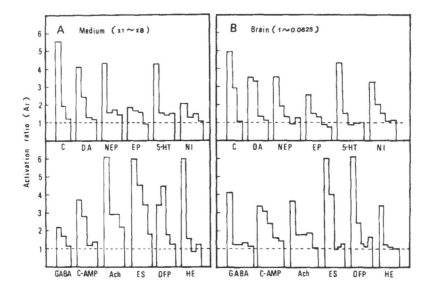

FIGURE 3.   The effect of various neurotransmitter candidates, their agonists, and antagonists on the *in vitro* release and accumulation of PTTH in pupal brain. Four separate cultures composed of four brains were each incubated in media containing various substances. Then the PTTH activity in the harvested medium (A) and in the extract (B) from cultured brains were measured for the 2× serial dilutions of samples using the *in vitro* PG assay. The left side of column in each treatment represents the original samples without any dilution. C represents control; DA, dopamine; NEP, norepinephrine; EP, epinephrine; 5-HT, serotonin; NI, nialamide; GABA, γ-aminobutyric acid; ACh, acetylcholine; ES, eserine; DFP, diisopropylphosphorofluoridate; HE, hemicolinium-3.

IV. DISCUSSION

It has not been clear whether transmitter substances have a direct or indirect effect on ecdysone synthesis by PG, although stimulation of ecdysone synthesis by PTTH requires extracellular calcium and cAMP as a second messenger for PTTH action on ecdysone synthesis.[21,22] Thus, when *in vitro* PG assays are used as an accurate measurement of PTTH activity in brain-cultured medium, which contains various transmitter substances, the evaluation concerning the effect of these substances on ecdysone synthesis by PG *in vitro* should be considered very carefully. The results obtained by treatments of different concentrations of transmitters on PG *in vitro* indicate that the *in vitro* PG assay for measuring PTTH activity is applicable for determination of PTTH activity in brain-cultured medium in the presence of neurotransmitters, since these chemicals (except the higher concentration of serotonin and nialamide) did not affect the basal ecdysone synthesis of day-0 *Mamestra* pupal PG *in vitro*.

There are three phenomena to consider in studies on release of a neurohormone from peptidergic neurosecretory systems:

1. Its synthesis in the neurosecretory cell body which is under the regulation of intrinsic and/or extrinsic stimuli
2. Its transport via axon tract to terminals
3. Its storage and release at axon terminals of corpora allata (CA), its neurohemal organ

Thus, each organ culture of the PTTH secretory system — brain-retrocerebral complex, isolated brain, or CA — is a suitable system to study the complicated regulation of PTTH release. In *Manduca*, when either the brain-retrocerebral complex or isolated CA as the PTTH neurohemal organ were subjected to culture, spontaneous release of large amounts of PTTH occurred.[13,15] The cultured day-0 *Mamestra* pupal brain also spontaneously released large amounts of PTTH into the culture medium during 24 h of incubation. However, cultured *Mamestra* brains with retrocerebral complexes released less PTTH from the neurohemal organ.[11] This result indicates that the neurosecretory cells in the brains of *Mamestra* pupae released PTTH via the cut end of the NCC I + II axonal tracts more easily than via the neurohemal organ, the CA. On the contrary, the brain-retrocerebral complexes of day-0 *Manduca* pupae released larger amounts of PTTH than isolated brain alone.[15] These apparent differences in PTTH release between two lepidopteran species, *Mamestra* and *Manduca*, could simply reflect interspecific variation and/or different culture conditions: in *Manduca*, the incubation period was 3 h and explants were subjected to culturing with part of the tracheal system attached; whereas, in *Mamestra*, explants were cultured for 24 h with no trachea. In order to investigate the regulation of PTTH synthesis in the soma of neurosecretory neurons, and transport in axonal tracts to terminals, cultures of the brain or neurohemal organs should be assessed separately. If the culture of brain-retrocerebral complexes is performed to clarify sites affecting transmitter substances on PTTH secretion, there may be complications in interpretation about action sites of neurotransmitter substances for the spontaneous PTTH release: the site is on neurosecretory soma and/or on axon terminals of the neurohemal organ.

The presence of biogenic amines in the insect nervous system has been reviewed fully.[5] However, in recent years, evidence has been presented which suggests a physiological relationship between aminergic and peptidergic neurons in insects. In particular, an association of biogenic amines with peptidergic neurosecretory systems has been emphasized by some authors, i.e., octopamine or synerphrine is mediating the release of adipokinetic hormone or hypertrehalosemic hormone in the grasshopper, *Locusta migratoria*,[23] and in the cockroach, *Periplaneta americana*;[24] and dopamine has elicited the release of myotropic ovulation hormone in the kissing bug, *Rhodnius prolixus*.[25] In the case of *Mamestra*, dopamine caused slight increase of PTTH release in culture, but epinephrine, norepinephrine, and the inhibitor of monoamine oxidase, nialamide, did not elicit PTTH release.

In some lepidopteran insects, it has been reported that large amounts of putative ACh is produced and retained in the brains of *Manduca*[26] and in the neurosecretory cells of brains in the European corn borer, *Ostrinia nubilalis*.[27] Interestingly, the cultured cells from the central nervous system of embryonic cockroaches contain neurons which are associated with ACh synthesis and physiologically functional ACh receptors.[28] In the present experiments, ACh as well as its agonists greatly stimulated PTTH release from *Mamestra* brains *in vitro*. This result suggests that PTTH release is a cholinergic-mediated neural system. In a peptidergic neurosecretory system, ACh may physiologically control the release of peptidergic neurohormones either by acting as a conventional neurotransmitter, regulating the activities of peptidergic neurosecretory cell bodies, or by acting as a neuromodulator, modifying the hormonal output from the peptidergic neurosecretory terminal. In the case of stimulation of PTTH release by ACh from isolated *Mamestra* brains, ACh may work at cholinergic neurons synapsed with the collaterals of PTTH cells, since the cultured explant in this experiment is lacking a neurosecretory terminal such as the neurohemal organ, the CA. In conclusion, organ culture of neurosecretory systems provides a unique preparation for detailed studies on the cholinergic system in insects, and, furthermore, represents additional information to support the view that ACh is a major neurotransmitter in the insect central nervous system, mediating the release of neurohormone from peptidergic neurosecretory cells.

ACKNOWLEDGMENTS

The author expresses his gratitude to Drs. L. I. Gilbert and W. E. Bollenbacher for supplying the ecdysone antibody, and further wishes to thank Dr. L. R. Whisenton for her critical reading of this manuscript. This work was supported by a grant-in-aid for special projects: Research on Mechanisms of Animal Behavior, from the Ministry of Education, Science, and Culture, Japan.

REFERENCES

1. **Scharrer, B. and Scharrer, E.**, Neurosecretion. VI. A comparison between the intercerebralis cardiacum-allata system of insects and the hypothalamo-hypophysial system of the vertebrates, *Biol. Bull.*, 87, 242, 1944.
2. **Hanstrom, B.**, Neurosecretory pathways in the head of crustaceans, insects and vertebrates, *Nature (London)*, 171, 72, 1953.
3. **McCann, S. M., Kalara, P. S., Donoso, A. O., Bishop, W., Schneider, H. P. G., Fawcett, C. P., and Krulich, L.**, The role of monoamine in the control of gonado-tropin and prolactin secretion, in *Proc. Int. Symp. Brain-Endocrine Interaction, Median Eminence*, Vandenhoeck and Ruprecht, Munich, 1971, 224.
4. **Weiner, R. I. and Ganong, W. E.**, Role of brain monoamines and histamines in regulation of anterior pituitary secretion, *Physiol. Rev.*, 58, 905, 1980.
5. **Evans, P. D.**, Biogenic amines in the insect nervous system, *Adv. Insect Physiol.*, 15, 317, 1980.
6. **Klemm, N.**, Histochemistry of putative transmitter substances in the insect brain, *Prog. Neurobiol.*, 7, 99, 1976.
7. **Kambysellis, M. and Williams, C. M.**, *In vitro* development of insect tissues. II. The role of ecdysone in the spermatogenesis of silkworms, *Biol. Bull. (Woods Hole Mass.)*, 141, 541, 1971.
8. **Agui, N.**, Activation of prothoracic glands by brain *in vitro*, *J. Insect Physiol.*, 26, 903, 1975.
9. **Bollenbacher, W. E., Agui, N., Granger, N. A., and Gilbert, L. I.**, *In vitro* activation of insect prothoracic glands by the prothoracicotropic hormone, *Proc. Natl. Acad. Sci. U.S.A.*, 76, 5148, 1979.

10. **Agui, N., Bollenbacher, W. E., and Gilbert, L. I.,** *In vitro* analysis of prothoracicotropic hormone specificity and prothoracic gland sensitivity in Lepidoptra, *Experientia,* 89, 984, 1983.

11. **Agui, N.,** *In vitro* secretion of prothoracicotropic hormone from the cultured brain of *Mamestra brassicae* L., in *Techniques in in Vitro Invertebrate Hormones and Genes,* Vol. C 209, Kurstak, E., Ed., Elsevier Scientific, County Clare, Ireland, 1986, 1.

12. **Bollenbacher, W. E., Agui, N., Granger, N. A., and Gilbert, L. I.,** Insect prothoracic glands *in vitro*: a system for studying the prothoracicotropic hormone, in *Invertebrate Systems in Vitro,* Kurstak, E., Maramorosch, K., and Dübendorfer, A., Eds., Elsevier/North-Holland, Amsterdam, 1980, chap. 22.

13. **Carrow, G. M., Calabrese, R. L., and Williams, C. M.,** Spontaneous and evoked release of prothoracicotropin from multiple neurohemal organs of the tobacco hornworm, *Proc. Natl. Acad. Sci. U.S.A.* 78, 5866, 1981.

14. **Bowen, M. F., Bollenbacher, W. E., and Gilbert, L. I.,** *In vitro* studies on role of the brain and prothoracic gland in the pupal diapause of *Manduca sexta, J. Exp. Biol.,* 108, 9, 1984.

15. **Bowen, M. F., Gilbert, L. I., and Bollenbacher, W. E.,** Endocrine control of insect diapause: an *in vitro* analysis, in *Techniques in in Vitro Invertebrate Hormones and Genes,* Vol. C210, Kurstak, E., Ed., Elsevier Scientific, County Clare, Ireland, 1986, 1.

16. **Agui, N., Ogura, N., and Okawara, M.,** Rearing of the cabbage armyworm, *Mamestra brassicae* L. (Lepidoptera: Noctuidae) and some lepidopterous larvae on artificial diets, *Jpn. J. Appl. Entomol. Zool.,* 19, 91, 1975.

17. **Yagi, S. and Honda, T.,** Endocrinological studies on pupal diapause on the cabbage armyworm, *Mamestra brassicae* L. I. Critical period on the brain hormones secretion for metamorphosis, *Jpn. J. Appl. Entomol. Zool.,* 21, 90, 1977.

18. **Horn, D. H. S., Wilkie, J. S., Sage, B. A., and O'Connor, J. D.,** A high affinity antiserum specific for the ecdysone nucleus, *J. Insect Physiol.,* 22, 901, 1976.

19. **Gilbert, L. I., Goodman, W., and Bollenbacher, W. E.,** Biochemistry of regulatory lipids and sterols in insects, in *International Review of Biochemistry: Biochemistry of Lipids II,* Goodwin, T. W., Ed., University Park Press, Baltimore, 1977, 1.

20. **Bollenbacher, W. E., Vedeckis, W. V., Gilbert, L. I., and O'Connor, J. D.,** Ecdysone titer and prothoracic gland activity during the larval-pupal development of *Manduca sexta, Dev. Biol.,* 44, 46, 1975.

21. **Smith, W. A., Gilbert, L. I., and Bollenbacher, W. E.,** The role of cyclic AMP in the regulation of ecdysone synthesis, *Mol. Cell. Endocrinol.,* 37, 285, 1984.

22. **Smith, W. A., Gilbert, L. I., and Bollenbacher, W. E.,** Calcium-cyclic AMP interactions in prothoracicotropic hormone stimulation of ecdysone synthesis, *Mol. Cell. Endocrinol.,* 39, 71, 1985.

23. **Orchard, I. and Loughton, B. G.,** Is octopamine a transmitter mediating hormone release in insect?, *J. Neurobiol.,* 12, 143, 1981.

24. **Downer, R. G. H., Orr, G. L., Gole, J. W. D., and Orchard, I.,** The role of octopamine and cyclic AMP in regulating hormone release from corpora cardiaca of the American cockroach, *J. Insect Physiol.,* 30, 457, 1984.

25. **Orchard, I., Ruegg, R. P., and Devey, K. G.,** The role of central aminergic neurones in the action of 20-hydroxyecdysone on neurosecretory cells of *Phodnius prolixus, J. Insect Physiol.,* 29, 387, 1984.

26. **Maxwell, G. D., Trait, J. F., and Hildebrand, J. G.,** Regional synthesis on neurotransmitter candidates in CNS of the moth *Manduca sexta, Comp. Biochem. Physiol.,* 61C, 109, 1979.

27. **Houk, E. J. and Beck, S. D.,** Distribution of putative neurotransmitter in the brain of the European corn borer, *Ostrinia nubilalis, J. Insect Physiol.,* 23, 1209, 1977.

28. **Beadle, D. J., Lees, G., and Bothan, R. P.,** Cholinergic neurones in neuronal cultures, in *Proc. Int. Symp. Insect Neurochemistry and Neurophysiology,* Borkovec, A. B. and Kelly, T. J., Eds., Plenum Press, New York, 1984, 317.

Chapter 13

EFFECTS OF TEST SUBSTANCES ON OVARIAN PROCESSES *IN VITRO* IN A PEDOGENETIC GALL MIDGE

D. F. Went and J. Kaiser

TABLE OF CONTENTS

I. INTRODUCTION

The gall midge *Heteropeza pygmaea* is a remarkable insect, as it can reproduce in the larval stage. This reproductive mode is called pedogenesis; it is linked to viviparity and parthenogenesis.[1] Thus, a young female larva develops into a so-called mother larva, from which, after a few days, young larvae will again hatch. As compared to oogenesis in other insects with polytrophic-meroistic ovarioles, oogenesis in pedogenetically reproducing *H. pygmaea* larvae is highly modified.[2] For example, true ovarioles are absent; the oocyte-nurse chamber complexes and, subsequently, the follicles are formed in a kind of ovarial sac. The follicles are released from the ovary simultaneously and develop in the hemolymph of the mother larva. Embryonic development in the follicles already starts in a developmental stage where in other dipteran insects vitellogenesis and oocyte growth only would set in.

In the last 10 years we have studied oogenesis in pedogenetic reproduction with a variety of techniques including light and electron microscopy, cinemicrography, X-ray irradiation, and indirect immunofluorescence.[3,4] Observation and manipulation of oogenetic processes have been highly facilitated by the elaboration of a method of culturing ovaries and follicles *in vitro*, i.e., in hemolymph obtained from sterilized larvae (summary in Went[4]). Oogenesis in this culture medium is normal, and it is thus possible to study the effects of hormones or drugs on ovarian development in an *in vitro* system. In the present report we will briefly summarize the effects of two selected test agents, i.e., 20-hydroxyecdysone and cytochalasin B, on specific oogenetic events, such as pulsation of oocyte nuclei, follicle formation, and synthetic ovarian activities.

II. MATERIAL AND METHODS

Larvae of *Heteropeza pygmaea* (Cecidomyiidae, Dipt.) were held under conditions where they reproduce by thelytokous pedogenesis only.[5] The methods used for culturing the ovaries *in vitro*, application of test agents, and time-lapse cinemicrography have been described elsewhere.[6-9] The radioactive precursors for RNA and protein synthesis we used were ³H-uridine and ³H-lysine (Amersham, England). After different periods of incubation the ovaries were washed several times with hemolymph and fixed and embedded according to Junquera.[10] Semithin sections were coated with Kodak® NTB-3 emulsion. Following exposure of 1 to 11 weeks the autoradiographs were developed with Kodak®-19. The sections were stained according to Musy et al.[11]

III. OOGENETIC EVENTS

At the time of explantation and start of the cultures the *H. pygmaea* ovary (60 to 80 μm in diameter) contains 15 to 20 oocyte-nurse chamber complexes and a large number of prefollicle cells. In the course of follicle formation, each oocyte-nurse chamber complex is surrounded by prefollicle cells, which thus constitute the follicular epithelium. This process takes 1 to 1.5 d. During the entire period of follicle formation the oocyte nuclei are seen to perform strong and continuous amoeboid movements which have been termed nuclear pulsations.[8,12] The significance of the motility of the oocyte nuclei is obscure, and it is also not yet clear whether this phenomenon occurs in ovaries of other insects as well.

³H-uridine given to the culture medium is clearly incorporated into the ovaries, indicating RNA synthesis (Figure 1). Some areas of the ovaries, indeed, contained no, or only a few, radioactive grains, suggesting a nonhomogeneous distribution of synthesis sites. However, the small size of the ovaries did not allow us to relate the sites of synthesis to specific ovarial components. Incorporation of ³H-lysine was also distinct, suggesting protein synthesis in the ovaries (Figure 2). Here, some variation in incorporation of radioactive precursors among the preparations was apparent.

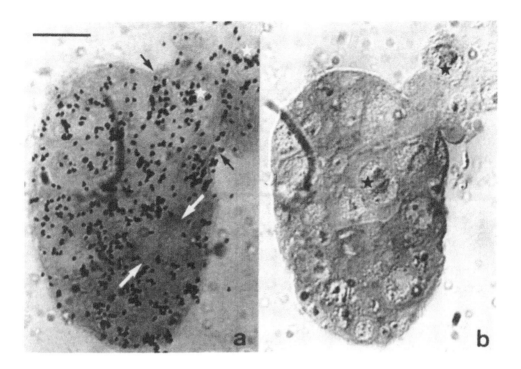

FIGURE 1. Ovary of *H. pygmaea*. (a) Autoradiograph showing distribution of labeled RNA after incubation of the ovary with ³H-uridine for 4 h. Some of the contents of the ovary had leaked out (between small arrows); this did not interfere with normal follicle formation. The large spherical nuclei (*) are nurse nuclei (cf. Went et al.[12]). Oocyte and prefollicle cell nuclei are less easy to identify in these sections. Note the area with fewer grains (large arrows). Bar = 20 μm; (b) ovary stained with methylene blue and safranin O (same ovary as in [a], but next section). Some nurse cell nuclei are clearly visible (*), but the identification of oocyte and prefollicle cell nuclei is still difficult.

IV. EFFECTS OF TEST SUBSTANCES

Pulsation of oocyte nuclei can be arrested within a few minutes by addition of low doses of cytochalasin B.[6] This reversible effect indicates that microfilaments are involved in the nuclear movements. On the other hand, the presence of cytochalasin B did not prevent follicle formation, although the nuclear pulsations were blocked. Similarly, neither the incorporation of radioactive uridine nor that of radioactive lysine into the ovaries seemed to be affected by addition of this drug to the cultures (cf. Figure 2).

The accelerating effect of 20-hydroxyecdysone on follicle formation in *H. pygmaea* ovaries cultured *in vitro* was already demonstrated several years ago.[9] A more recent and detailed analysis shows that this hormone successively stimulates migration and aggregation activity of the prefollicle cells.[13] In contrast, 20-hydroxyecdysone was not seen to exert any effects (stimulating or retarding) on the pulsations of the oocyte nuclei. Possible interactions between the molting hormone and synthetic activities in the ovary have not yet been studied.

V. DISCUSSION

Many workers on insect oogenesis have studied the synthetic capacities of the ovary and its components (review in Nardon[14]). In insects with polytrophic-meroistic ovarioles the stage of oocyte growth is usually characterized by strong RNA synthesis. In particular, the nurse chamber shows rapid incorporation of labeled precursors into RNA. The oocyte nucleus

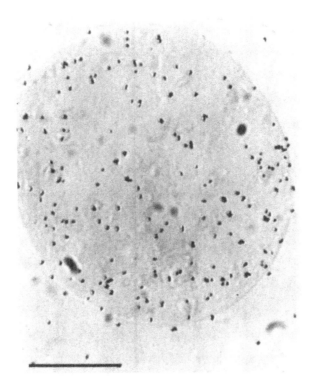

FIGURE 2.   Ovary of *H. pygmaea*. Autoradiograph showing incorporation of ³H-lysine after incubation of the ovary for 1 h. Cytochalasin
B was given to this culture in a final concentration of 14 μg/ml culture
medium 15 min before ³H-lysine was added. Bar = 20 μm.

itself is usually synthetically inactive, but exceptions concerning specific oogenetic stages
or species may occur. Such an exception seems to be the (nonpedogenetic) gall midge
Wachtliella persicariae.[15] Most of the proteins accumulating in the oocyte are exogenous
in origin, but at least some proteins are synthesized inside the ovary (cf. Nardon[14]). The
autoradiographic data given in this report indicate that both RNA and protein synthesis occur
in the *H. pygmaea* ovary as well. This study differs, however, from most other autoradiographic studies of insect oogenesis in several respects. First, we used an *in vitro* system,
so that we can presume that the sites of labeling also represent the sites of synthesis. Second,
the mode of oogenesis under study is modified and adapted to pedogenetic reproduction.
Third, the stage of oogenesis we examined, namely the stage of follicle formation, precedes
the oogenetic stage usually studied, i.e., vitellogenesis, considerably. It is unfortunate,
however, that due to the small dimensions of the chosen objects, the results do not allow
us to give more precise information. Thus, we do not know if all cell types in the ovary
(oocytes, nurse cells, and prefollicle cells) are synthetically active or only the nurse cells.
It should also be mentioned here that these results are still preliminary because some control
experiments have to be carried out yet.

The oogenetic events examined in this report (nuclear pulsation, follicle formation, and
macromolecular synthesis) occur simultaneously. Based on this temporal coincidence, interactions between these processes might be presumed. The present data allow us (with some
restrictions) to rule out direct connections between the pulsating oocyte nuclei, on the one
hand, and follicle formation and synthetic activities in the ovary, on the other hand. While
direct connections do not seem to exist, indirect connections are still conceivable. It is, for
example, possible that the nuclear movements of the oocytes enable or facilitate the transport

of synthesized products through the nuclear envelope. Inhibition of nuclear movements during follicle formation may then interfere with normal follicle growth and development only in a later stage of oogenesis. To investigate possible relations between pulsation of oocyte nuclei and penetrability of the nuclear envelopes as well as between hormone treatment and synthetic activity, it will be necessary to combine the methods of *in vitro* culturing with that of autoradiography and electron microscopy. We hope that by a sophisticated application of modern techniques, in the end we will be able to answer some of the questions posed by the extraordinary mode of oogenesis of the pedogenetic insect *H. pygmaea*.

ACKNOWLEDGMENTS

This work was supported by Swiss National Science Foundation Grant No. 3.193-0.82.

REFERENCES

1. **Camenzind, R.**, Untersuchungen über die bisexuelle Fortpflanzung einer paedogenetischen Gallmücke, *Rev. Suisse Zool.*, 69, 377, 1962.
2. **Went, D. F.**, Paedogenesis in the dipteran insect *Heteropeza pygmaea*: an interpretation, *Int. J. Invertebr. Reprod.*, 1, 21, 1979.
3. **Junquera, P.**, Oogenesis in a paedogenetic dipteran insect under normal conditions and after experimental elimination of the follicular epithelium, *Roux's Arch. Dev. Biol.*, 193, 197, 1984.
4. **Went, D. F.**, Insect ovaries and follicles in culture: oocyte and early embryonic development in pedogenetic gall midges, *Adv. Cell Cult.*, 2, 197, 1982.
5. **Went, D. F.**, Role of food quality versus quantity in determining the developmental fate of a gall midge larva (*Heteropeza pygmaea*) and the sex of its paedogenetically-produced eggs, *Experientia*, 31, 1033, 1975.
6. **Kaiser, J., Lang, A. B., and Went, D. F.**, Pulsation of nuclei in insect oocytes is reversibly inhibited by cytochalasin B, *Exp. Cell Res.*, 139, 460, 1982.
7. **Went, D. F.**, *In vitro* culture of ovaries of a viviparous gall midge, *In Vitro*, 13, 76, 1977.
8. **Went, D. F.**, Pulsating oocytes and rotating follicles in an insect ovary, *Dev. Biol.*, 55, 392, 1977.
9. **Went, D. F.**, Ecdysone stimulates and juvenile hormone inhibits follicle formation in a gall midge ovary *in vitro*, *J. Insect Physiol.*, 24, 53, 1978.
10. **Junquera, P.**, Polar plasm and pole cell formation in an insect egg developing with or without follicular epithelium: an ultrastructural study, *J. Exp. Zool.*, 227, 441, 1983.
11. **Musy, J. P., Modis, L., Gotzos, V., and Conti, G.**, Nouvelles méthodes de coloration sur coupes semifines pour tissus inclus en "Araldit". Etudes au microscope à champ clair, à contraste de phase et à fluorescence, *Acta Anat.*, 77, 37, 1970.
12. **Went, D. F., Fux, T., and Camenzind, R.**, Movement pattern and ultrastructure of pulsating oocyte nuclei of the paedogenetic gall midge, *Heteropeza pygmaea* Winnertz (Diptera: Cecidomyiidae), *Int. J. Insect Morphol. Embryol.*, 7, 301, 1978.
13. **Went, D. F.**, Effects of 20-hydroxyecdysone on cultured ovaries of a pedogenetic gall midge, in *Techniques in in Vitro Invertebrate Hormones and Genes*, Vol. C204, Kurstak, E., Ed., Elsevier, County Clare, Ireland, 1986, 1.
14. **Nardon, P.**, La synthèse et l'accumulation d'acides nucléiques et de protéines au cours de l'ovogenèse chez les insectes, *Ann. Biol.*, 17, 105, 1978.
15. **Kunz, W., Trepte, H.-H., and Bier, K.**, On the function of the germ line chromosomes in the oogenesis of *Wachtliella persicariae* (Cecidomyiidae), *Chromosoma*, 30, 180, 1970.

Chapter 14

HORMONAL CONTROL OF CUTICULAR MELANIZATION *IN VITRO*

Kiyoshi Hiruma and Lynn M. Riddiford

TABLE OF CONTENTS

I. INTRODUCTION

Many insect species show "morphological color change" during their lives due mostly to the synthesis of melanin, ommochrome, and other pigments.[1,2] Most of these changes are mediated by the endocrine system during a molt and are based on a slow and long-lasting process that involves a change in the amount of pigment and/or of pigment cells. Yet, very little is known about the action of the hormones on pigment synthesis at the cellular and the biochemical levels.

In the tobacco hornworm, *Manduca sexta*, when juvenile hormone (JH) is absent at the critical period around the time of head capsule slippage (HCS) during the larval molt, the newly synthesized cuticle melanizes shortly before ecdysis.[3-5] Here we summarize recent *in vitro* studies of the hormonal regulation of this melanization and associated enzymes, phenoloxidase (PO) and dopa decarboxylase (DDC).

II. HORMONAL CONTROL OF CUTICULAR MELANIZATION

Normally, *Manduca* larvae have a transparent cuticle. During a larval molt the absence of JH around the time of HCS (near the peak of the ecdysteroid titer) causes the deposition of premelanin granules containing an inactive pro-phenoloxidase into the newly forming cuticle 13 h later. When the ecdysteroid titer declines (Figure 1A), the enzyme is activated, and melanization occurs 3 h before ecdysis (26 h after HCS) within the granules.[3,6]

Although integument, explanted 15 or more hours after HCS, melanized *in vitro*[7] (see Section III below), that explanted before HCS formed only a transparent new cuticle which lacked premelanin granules.[8] These findings led to the discovery of an abdominal factor that is also necessary for deposition of premelanin granules.[9] Ligation experiments showed that this factor comes from the fifth abdominal segment and is apparently released some time in the 5 h before HCS (Figure 1A). Nerve cord homogenates can substitute for this factor when injected into anterior portions of larvae ligated between the third and fourth abdominal segments and are now being tested for their ability to cause premelanin granule deposition *in vitro*.

III. MELANIZATION *IN VITRO* AND THE DISCOVERY OF DOPAMINE MELANIN

Integumental pieces from allatectomized larvae, which were explanted 10, 6, and 3 h before ecdysis and cultured in β-alanine-free Grace's medium, melanized.[7] Importantly, consistent melanization required that the pieces be placed on glass wool to ensure their position at the surface and, thus, access to maximal oxygen.

Success of the *in vitro* melanization allowed us to determine that this melanin was a dopamine melanin; heretofore, most insect melanins have been considered to be indole eumelanin.[10,11] Addition of 0.3 mM of either dopa or dopamine improved the melanization *in vitro* so that it appeared as *in vivo*.[7] When the enzyme DDC which converts dopa to dopamine was inhibited, melanization *in vitro* occurred only in the presence of dopamine.[6] Furthermore, ^{14}C-dopamine was incorporated into the melanin *in vitro*[7] and *in vivo*.[12] Also, the solubility of the melanin in sodium hydroxide[13] indicated dopamine melanin.[6] Therefore, both PO and DDC are key enzymes in the melanization process in *Manduca* cuticle.

IV. DOPA DECARBOXYLASE (DDC)

DDC activity was found primarily in the epidermis with small amounts in the cuticle and the fat body and none in the hemolymph.[6] The epidermal DDC activity was low until 12 h

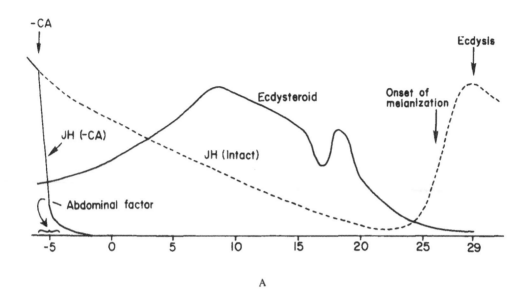

A

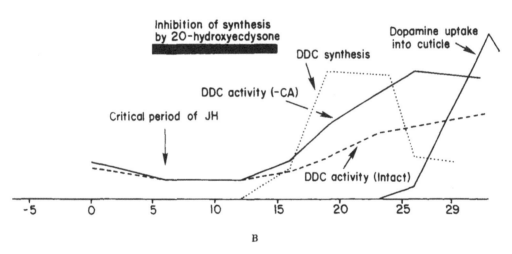

B

FIGURE 1. (A) Ecdysteroid titer, JH titer, and timing of abdominal factor release during the last larval molt in *Manduca*; -CA indicates allatectomy. Data for ecdysteroid titer, JH titer, and abdominal factor are from Curtis et al.,[3] Fain and Riddiford,[18] and Hiruma and Riddiford,[9] respectively. (B) Temporal events of epidermal dopa decarboxylase (DDC) synthesis[8] and activity,[14] and timing of uptake of dopamine into the cuticle[6] during the final larval molt. All measurements were made on allatectomized larvae except for DDC activity in intact larvae. (C) Timing of phenoloxidase (PO) synthesis, deposition, and activation during the final larval molt of allatectomized larvae. The data for deposition of premelanin granules and activation of the phenoloxidase are from Curtis et al.[3] and Hiruma et al.[6] The synthesis and the titer of PO are based on studies with the PO antibody.[17]

after HCS, then increased to its maximal level at the onset of melanization[6,14] (Figure 1B). This activity was precipitated by the antibody to *Drosophila* DDC, and pulse-chase experiments *in vitro* showed that the epidermis synthesized a 49-kDa polypeptide, precipitable with the antibody during the period that DDC activity was increasing[8] (Figure 1B). By the onset of melanization, dopamine was accumulating in the epidermis; then as melanization proceeded, dopamine was incorporated into the cuticle[6] (Figure 1B).

The hormonal control of DDC synthesis is complex. The absence of JH during the high ecdysteroid titer just after HCS causes a twofold higher maximal activity at the onset of

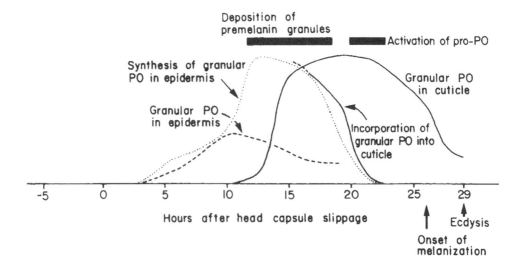

FIGURE 1C

melanization.[14] However, DDC synthesis does not begin until 16 h later (Figure 1B), and this onset of synthesis can be inhibited by 20-hydroxyecdysone (20-HE), both *in vivo* and *in vitro*.[14,15] Moreover, *in vitro*, 20-HE was most effective if given before the onset of synthesis,[14] indicating that 20-HE may be interfering with the production of the DDC mRNA. Thus, the determination of later DDC expression appears to be during the rise of ecdysteroid, at which time it can be modulated by JH, whereas the timing of the synthesis itself is governed by the decline of ecdysteroid.

V. PHENOLOXIDASE (PO)

The PO in the premelanin granules is a diphenoloxidase and was found to be separable by electrophoresis from both the wound PO[16] found in the old cuticle and the hemolymph PO, so it was isolated and an antibody prepared.[17] A ^{3}H-leucine pulse-chase *in vitro* followed by precipitation with the PO antibody showed that the epidermis synthesized this granular PO, beginning about 6 to 10 h after HCS (Figure 1C). Maximal synthesis of the PO occurred 12 to 16 h after HCS during which time PO appeared in the cuticle. Cessation of epidermal synthesis and cuticular deposition of the PO occurred about 21 to 23 h after HCS (Figure 1C). This timing coincides well with the timing of deposition of the premelanin granules.[3,6]

During synthesis and deposition the PO is inactive. It then becomes activated about 21 to 23 h after HCS.[6] *In vitro* as well as *in vivo*, 20-HE prevented the activation, indicating that activation occurred due to epidermal events. Preliminary experiments indicate that a new serine protease appears in the new cuticle at this time,[8] but its relationship to the activation of the granular PO has not been determined. Once activated, this PO can oxidize the dopamine — synthesized by the epidermis and transported to the granules — to melanin within the granules. As melanin deposition proceeds, the PO becomes less active (Figure 1C).

VI. CONCLUSION

The use of *in vitro* culture has allowed us to elucidate the hormonal regulation of melanization in *Manduca* larvae. An abdominal neurosecretory factor is necessary for the deposition of the premelanin granules in the absence of JH. Ecdysteroid and JH regulate synthesis and/

or activation of two key enzymes necessary for the melanization, so that they ultimately control the melanization processes. These enzymes, PO and DDC, are good candidates for the study of the molecular basis of the action of JH in the determination of melanization and of the fall of ecdysteroid in the onset of melanization.

ACKNOWLEDGMENTS

We thank Dr. Ross Hodgetts for a gift of the *Drosophila* DDC antibody. The unpublished studies have been supported by NSF PCM80-11152 and PCM85-10875 and NIH AI12459.

REFERENCES

1. **Raabe, M.,** *Insect Neurohormones,* Plenum Press, New York, 1982.
2. **Riddiford, L. M.,** Hormone action at the cellular level, in *Comprehensive Insect Physiology Biochemistry and Pharmacology,* Vol. 8, Kerkut, G. A. and Gilbert, L. I., Eds., Pergamon Press, Oxford, 1985, 37.
3. **Curtis, A. T., Hori, M., Green, J. M., Wolfgang, W. J., Hiruma, K., and Riddiford, L. M.,** Ecdysteroid regulation of the onset of cuticular melanization in allatectomized and *black* mutant *Manduca sexta* larvae, *J. Insect Physiol.,* 30, 597, 1984.
4. **Truman, J. W., Riddiford, L. M., and Safranek, L.,** Hormonal control of cuticle coloration in the tobacco hornworm: basis of ultrasensitive assay for juvenile hormone, *J. Insect Physiol.,* 19, 195, 1973.
5. **Riddiford, L. M. and Hiruma, K.,** Regulation of melanization in insect cuticle, in *Advances in Pigment Cell Research,* Bagnara, J. T., Ed., Alan R. Liss, New York, 1988, 423.
6. **Hiruma, K., Riddiford, L. M., Hopkins, T. L., and Morgan, T. D.,** Roles of dopa decarboxylase and phenoloxidase in the melanization of the tobacco hornworm and their control by 20-hydroxyecdysone, *J. Comp. Physiol.,* 155B, 659, 1985.
7. **Hiruma, K. and Riddiford, L. M.,** Regulation of melanization of tobacco hornworm larval cuticle *in vitro, J. Exp. Zool.,* 230, 393, 1984.
8. **Hiruma, K. and Riddiford, L. M.,** unpublished data, 1987.
9. **Hiruma, K. and Riddiford, L. M.,** Hormonal regulation of the deposition of premelanin granules and of the enzymes involved in melanization of *Manduca sexta, Abstr. 4th Int. Symp. of Juvenile Hormones,* Niagara-On-the-Lake, Ontario, 1986, 20.
10. **Hackman, R. H.,** Chemistry of insect cuticle, in *The Physiology of Insecta,* Vol. 6, Rockstein, M., Ed., Academic Press, New York, 1974, 216.
11. **Needham, A. E.,** Insect biochromes, in *Biochemistry of Insects,* Rockstein, M., Ed., Academic Press, New York, 1978, 233.
12. **Hori, M., Hiruma, K., and Riddiford, L. M.,** Cuticular melanization in the tobacco hornworm larva, *Insect Biochem.,* 14, 267, 1984.
13. **Swan, G. M.,** Structure, chemistry, and biosynthesis of the melanins, *Fortschr. Chem. Org. Naturst.,* 31, 521, 1974.
14. **Hiruma, K. and Riddiford, L. M.,** Hormonal regulation of dopa decarboxylase during a larval molt, *Dev. Biol.,* 110, 509, 1985.
15. **Hiruma, K. and Riddiford, L. M.,** Inhibition of dopa decarboxylase synthesis by 20-hydroxyecdysone during the last larval molt of *Manduca sexta, Insect Biochem.,* 16, 225, 1986.
16. **Barrett, F. M.,** Wound-healing phenoloxidase in larval cuticle of *Calpodes ethlius* (Lepidoptera: Hesperiidae), *Can. J. Zool.,* 62, 834, 1984.
17. **Hiruma, K. and Riddiford, L. M.,** Granular phenoloxidase involved in cuticular melanization in the tobacco hornworm: regulation of its synthesis in the epidermis by juvenile hormone, *Dev. Biol.,* in press.
18. **Fain, M. J. and Riddiford, L. M.,** Juvenile hormone titers in the hemolymph during late larval development of the tobacco hornworm, *Manduca sexta* (L), *Biol. Bull.,* 149, 506, 1975.

*Applications of Insect Cell Cultures in
Biotechnology and Molecular Biology*

Chapter 15

CELL FUSION STUDIES ON INVERTEBRATE CELLS *IN VITRO*

H. M. Mazzone

TABLE OF CONTENTS

I. INTRODUCTION

Approximately 30 years ago in Japan, Okada and co-workers[1] demonstrated that the efficiency of cell fusion *in vitro* was enhanced by a parainfluenza virus, the hemagglutinating virus of Japan. The virus was first isolated in Sendai at the Tohuku University School of Medicine by Kuroya et al. in 1953,[2] and subsequently designated as HVJ by the Society of Japanese Virologists. In the West, Sendai virus is a synonym commonly used for this microorganism.

During the initial development of somatic cell genetics by Barski et al.[3] and the extension of such work by Sorieul and Ephrussi[4], hybrids were formed without the aid of exogenously added virus. The early work fostered the impression that cells in culture naturally maintained a potentiality for fusion and that fusion between cells of different species could occur.

However, fusion occurring within a given cell culture can be a rare event. From a number of studies (cf. Harris[5]), the efficiency of hybrid formation was found to increase about 1000-fold when fusion was prompted by a virus. Another important consideration was that the use of a virus to achieve cell fusion did extend the range of cell types that form hybrids.[6] The phenomenon of virus-induced cell fusion became a standard method for the formulation of heterokaryons and for somatic hybrid formations using, in many cases, attenuated preparations of HVJ (Sendai virus).

Agents that promote cell fusion, fusogens, are classified as biological, chemical, or physical. This report discusses all three types of cell fusion techniques and cites the biotechnological potential of cell fusion experiments. Finally, particular applications of fusion techniques to invertebrate cells in culture are presented.

II. FUSION-INDUCING AGENTS AND ACCOMPLISHMENTS DERIVED FROM THEIR USE

A. Introduction

As examples of biological fusogens, a number of viruses are reported to have this capability (cf. Ringertz and Savage[7]):

1. Strains of herpes — varicella, herpes simplex
2. Rabbitpox
3. Paramyxoviruses (in addition to HVJ) — mumps, Newcastle disease, SV 5, measles, respiratory syncytia
4. Oncornaviruses — Rous sarcoma, Visna
5. Coronavirus — avian infectious bronchitis

Chemical fusogens include polyethylene glycol (PEG), concanavalin A (Con A), and wheat germ agglutinin. Physical agents that promote fusion of cells are modulated electric pulses and laser light. We will briefly review some major accomplishments resulting from the three types of fusion-promoting agents.

B. Biological Fusion Agents

Hybrid formation of cells through use of a biological agent has reached its greatest impact, thus far, in the work of Köhler and Milstein in 1975.[8] Immunized mouse cells were fused to cancerous mouse cells through the action of attenuated HVJ. Continuous cultures of hybrids were obtained, capable of secreting antibodies of predefined specificity, i.e., monoclonal antibodies. The resulting hybridomas inherited two vital traits from the parental cells: from the immunized blood cell, the ability to produce the specific antibody required; and from the cancer cell, the ability to replicate endlessly. I believe that the second trait has been overshadowed by the first trait, and, in this connection, researchers should realize the potential of cell fusion to aid in a variety of studies on invertebrate cell cultures.

C. Chemical Fusion Agents

In 1975 interkingdom interblending of cells was accomplished.[9] The cell walls of plant cells, fungi, and bacteria inhibit somatic cell fusion. It was not until wall-less plant, fungal, and bacterial protoplasts became available (cf. Cocking[10]) that this capability could be extended. By 1975, PEG had been introduced as a chemical fusogen (cf. Ringertz and Savage[7]). It was used to induce the fusion of protoplasts and also the fusion of hen erythrocytes. The availability of this common chemical agent, which could be used effectively for the fusion of both animal cells and protoplasts, led to an investigation that PEG might likewise induce fusion between hen erythrocytes and yeast protoplasts. Fusion was observed, and it was suggested that such investigation might further our knowledge of nuclear and cytoplasmic interactions in animal-fungal heterokaryons.[9]

A method employing high pH, high concentration of calcium ions, and raised temperature was reported to be more effective in fusing plant and animal cells than the PEG procedure. Enzyme treatment (protease) also increased the efficiency observed in these latter experiments.[11]

D. Physical Fusion Agents

The technique of electrofusion (cf. Zimmerman et al.[12]) involves the establishment of membrane contact between cells by exposing them to an alternating (nonhomogeneous) electric field. This condition leads to dielectrophoresis — the movement of a neutral particle in the direction of the net force acting on the particle in a nonuniform alternating current (AC) field. As a result, cell contact is made in the form of chain-like aggregates, resembling strings of pearls. Once cell-to-cell contact is established, fusion is induced by a high-voltage direct current (DC) pulse of microsecond duration. The membrane undergoes physical breakdown leading to the formation of minute pores, and cell fusion proceeds as a result of the membrane breakdown in areas of cell-to-cell contact.

A modification of the above procedure avoids long exposure of cells to alternating fields. Attention is directed to the application of a homogeneous electric field (DC) of relatively short duration to result in pore formation in the cell membranes, electroporation. However, as noted above, homogeneous electric field pulses cause membrane-to-membrane fusion when cells are in close contact (cf. Neuman and Bierth[13]).

The development of the electric field-induced fusion of cells appears to have certain advantages over chemical- or virus-induced fusion of cells. These advantages include: (1) less cell damage, (2) higher yield of fused cells, and (3) greater control over the fusion process. Chemical- and viral-induced cell fusions proceed in an asynchronous manner. In asynchronous fusion, cells may cluster through the fusion process; however, for a hybrid to be capable of division, it is vital that the nuclei of the two original cells are in the correct phase for genetic exchange to take place. This accomplishment will not necessarily be guaranteed if asynchronous cell cultures are used, so that fusion will result in cell union but not in nuclear exchange. In electrofusion, the probability of genetic exchange taking place increases considerably when fusion occurs between more than one of each of the respective original cells. This situation is more often the case in electrofusion than in biological or chemical fusion methods.

Another physical fusion technique involves a laser microbeam introduced into a light microscope[14] where a beam splitter conducts the microbeam down through the objective onto a culture of cells. A vacuum pump generates a permanent N_2 gas flow in the laser tube of a nitrogen laser. At high voltages of 10 to 16 kV, very short light pulses are emitted with a wavelength of 337 nm. These pulses, in turn, trigger the emission of light from a dye

Table 1
SOME BIOTECHNOLOGICAL POSSIBILITIES UTILIZING CELL FUSION PROCEDURES

Experiment	Ref.
Chromosomal localization of genes modeled after classical experiments	Weiss and Green (1967)[33]
Improved production of monoclonal antibodies	Zimmerman and Greyson (1983)[41]
Low copy number transfection of cells	cf. Neumann and Bierth (1986)[13]
Transformation of yeast cells	Hashimoto et al. (1985) after Neumann and Bierth[13]
Exchange of genetic material among multifused cells	Zimmerman and Vienken (1982),[42] Neumann and Bierth (1986)[13]
Fusion of cell organelles	Zimmerman and Vienken (1982)[42]
Fusion of drug- and metabolite-containing lipid vesicles with living cells	cf. Kuta et al. (1985)[39]
Hybrids capable of fixing atmospheric N_2	Poste and Nicholson (1978)[43]
Tolerance to salt of crops through fusion with marine algae or bacteria	Zimmerman and Steudle (1978)[44]
Plant cell modification leading to improved crops	Power et al (1978),[45] Ringertz and Savage (1976)[7]

Adapted in part from Kuta et al.[39]

laser, and, depending on the dye, the laser beam can vary in wavelength from approximately 365 to 870 nm. Parallel to the beam of the dye laser is that of a red helium-neon pilot laser, used to visualize the target area that will be hit by the dye laser. The laser technique permits exclusive fusion of two, preselected, adjacent cells within an aggregate of cells at precisely chosen points of development and cell cycle phase.

From the above discussion it should be clear that interested investigators have available the full range of cell fusion techniques in terms of utilizing biological, chemical, and physical agents. Indeed, at times it may be advantageous to combine one method with another as demonstrated in an electrochemical fusion technique by Weber and co-workers.[15,16]

III. BIOTECHNOLOGICAL POTENTIAL OF CELL FUSION PROCEDURES

Perhaps of greatest interest to cell experimenters is that fusion procedures allow for direct transfer of genetic material between fused cells. Auer et al.,[17] employing an electrofusion technique, found uptake of viral DNA from SV 40 and RNA from mammalian cells in human red blood cells. Moreover, fusion effects on membrane structures facilitate, in many cases, the transfer of macromolecules or larger particles into the interior of the cell.[18]

The researcher, and particularly the cell researcher, may draw on classical experiments as well as present developments in the application of cell fusion procedures. In this connection, Table 1 cites several types of experiments where cell fusion research may produce a biotechnological result. Obviously the table is incomplete, and some experiments, as they are being applied to cell fusion research, are only in the beginning stages of development.

IV. APPLICATION OF CELL FUSION TECHNIQUES TO INVERTEBRATE CELLS *IN VITRO*

Earlier attempts to fuse invertebrate cells were mostly limited to cells of *Drosophila* species. Evidence that *Drosophila* cells are capable of fusing spontaneously in *in vitro* preparations comes from the work of Barigozzi.[19] In the study, analysis indicated extra chromosomes, particularly the X chromosome, within certain cells.

Artificial fusion of *Drosophila* cells with a number of agents has been tried by several investigators. Thus far, biologically induced fusion using HVJ (Sendai virus) has not been achieved with *Drosophila* cells. This condition could be the result of a lack of terminal sialic sites which serve as receptors for the paramyxoviruses, such as HVJ.[20] Chemically induced fusion of *Drosophila* cells with Con A, wheat germ agglutinin, lysolecithin, and PEG have been reported.[21-24] In these reports, the fusion capability of the chemicals was demonstrated by autoradiographs of dikaryons with the presence of one labeled and one unlabeled nucleus. However, the fate of the hybrid cells was not followed. Successful fusion (using a chemical agent) and hybrid proliferation were reported for *Drosophila* cells by Nakajima and Miyake[20] and by Wyss.[25] In each case, PEG was used as the fusing agent, and in the study by Nakajima and Miyake, dimethyl sulfoxide enhanced the fusion process. The PEG exposure to the *Drosophila* cells was of brief duration, 30 s to 1 min, for each study. Mitsuhashi[26] reported on *in vitro* attempts to fuse cells from mosquito, fly, beetle, and moth insects using biological (HVJ) and chemical (PEG, Con A) agents. The use of PEG resulted in a low rate of fusion of mosquito cells and of fly cells. Interspecific cell fusion was rare. Preliminary electrofusion experiments indicated that each of the cell lines might require specific conditions for the successful fusion of cells.

An important application of cell fusion techniques should be directed to the problems of pest insects and crop diseases. In the case of pest insects, cells from established lines of a given insect could be fused to cells from other organs of the same insect. This possibility has been demonstrated in the case of a beneficial insect, *Drosophila*, by Wyss.[27] Cells from an established ecdysone-sensitive cell line (MDR 3) were fused by PEG to cells derived from different developmental stages of the insect. Alternatively, the cells of the established cell line could be fused to cells from a different insect, i.e., interspecies fusion.

In this research, it would also be worthwhile to repeat the Köhler-Milstein experiment.[8] One component of the hybrid would be a tumorous or cancerous cell, obtained either from insects[28] or from other phyla, e.g., HeLa cells (derived from a human carcinoma). In this connection, we should note that the feasibility of such experimentation was demonstrated by Johnson and co-workers[29] and Conover and co-workers.[30] In these studies, interphylum hybrids were obtained involving the fusion of human cancer cells (HeLa) and insect cells. The fusing agent was UV inactivated HVJ (Sendai virus).

Such studies are quite appropriate to invertebrate cells *in vitro*. The feasibility is significant in regard to the use of viruses as biological insecticides. For example, at present we have no cell line derived from the European pine sawfly (*Neodiprion sertifer*, Geoffroy), an insect pest of evergreen trees. The Baculovirus isolated from it can be purified in high quantities[31] and is very effective in the field against host larvae.[32] However, in order to study virus-cell interactions, we are limited to doing so when the insect is available yearly, from May to June.

When cells are fused, usually the chromosomes of one or the other "parent" of a hybrid cell are gradually lost over successive generations.[33] Pontecorvo[34] demonstrated that X-irradiation, gamma irradiation, or deoxybromouridine (BUDR) labeling of one "parent" before fusion could be used to predetermine which of the two sets of chromosomes would be, by choice of the investigator, preferentially lost. The loss of one species' chromosomes from a hybrid results in chromosome segregation. This condition allows for the association of genetic function with specific chromosomes in eukaryotic cells. For example, by analyzing the hybrid for protein markers of the cells losing chromosomes, it is possible to locate the structural genes for the proteins on that specific chromosome. This is an area of research that would provide insight into the *biology* of pest insects, an area that becomes more significant as such insects become more and more resistant to current insecticides, both chemical and biological.

Much of the work in our laboratory has been concerned with the gypsy moth. While we have registered a Baculovirus insecticide with the U.S. Environmental Protection Agency, it is not very effective. The insect is spreading, and a number of research programs are being considered to increase the efficacy of the virus. One consideration involves a study of latent viruses in insects.[35] We have experienced that primary cultures of gypsy moth hemocytes can be maintained for long periods so that they grow almost as monolayers.[36,37] Frequently, such cells show spontaneous virus development. We would like to exploit this event in the following manner.

In some cell systems it has been reported that cell fusion can be used as a means for the detection of latent viruses. It has been shown that certain virus-induced mouse tumors consist of cells which multiply for many cell generations, show malignant properties, and yet do not contain any detectable quantities of tumor virus. However, when some of these tumor cells are fused with normal cells that permit virus multiplication, the resulting hybrid cells produce a virus. The confrontation of the tumor cell nucleus with the cytoplasm of the sensitive cell somehow awakens ("rescues") a dormant virus in the tumor nucleus.[7] In this way, cell fusion could be used as an analytical tool for screening gypsy moth egg masses for the presence of virus. Egg masses would be collected before the hatching season, and the larvae, reared and bled to obtain hemocytes. The hemocytes would be fused with appropriate cells to express latent viruses. Knowing the extent of virus would help in our strategy for spraying virus insecticide in insect-infested areas. We are doing preliminary studies in this regard. Apparently, gypsy moth hemocytes can be fused by various techniques cited in this report, especially by the relatively new procedure of electrofusion.

Cell fusion techniques may also be applied to environmental problems involving microbial insecticides and fungus-caused tree diseases. Shivarova and Grigorova, using electrofusion procedures, were able to successfully fuse protoplasts of the bacterial insecticide, *Bacillus thuringiensis*, containing a plasmid, pUB 110, which codes for kanamycin resistance, with kanamycin-sensitive protoplasts lacking pUB 110. The resulting homokaryons were kanamycin resistant and stable after 20 passages.[38] Such methodology could be applied to the fusion of *B. thuringiensis* to other bacteria such as species of *Pseudomonas* which have better environmental qualities than *B. thuringiensis*.

Schnettler et al. in 1984 (cf. Shivarova and Grigorova[38]) achieved good success using electrofusion to generate yeast hybrids. In these experiments, protoplasts of auxotrophic mutants of *Saccharomyces cerevesiae* were fused. The *S. cerevesiae* protoplasts containing the plasmid, paDh 040-2, which carries the β-lactamase gene, were fused with protoplasts lacking this plasmid. Hybrids grew in minimal medium, which permitted only growth of fusion products, and also expressed the β-lactamase gene. A higher yield of hybrids was generated using electrofusion procedures over other types of fusing agents.

A similar hybridization effect could be realized in fusing aggressive strains of fungi which cause tree diseases, e.g., Dutch elm disease and chestnut blight, with nonaggressive strains of these fungi. Such results would greatly increase the frequency of aggressive-nonaggressive hybrids over what is observed in nature. Cloning for the hybrid favoring nonaggressiveness would produce a fungal hybrid that could override the effect of aggressive fungal infection in elms and chestnuts.

Shimizu[40] investigated protoplast fusion by PEG of the insect pathogenic fungi, *Beauveria bassiana*, *B. brongniartii*, and *B. amorpha*. Intraspecific protoplast fusion of *B. bassiana* and *B. amorpha* produced prototrophic colonies of high frequencies. Isolates of *B. brongniartii* were classified into two types, type 1 and type 2, based on their electrophoretic patterns of certain enzymes. In the fusion of protoplasts of the same type, high frequencies were obtained, whereas in the protoplast fusion between type 1 and type 2, the complementation frequency was low. In interspecific protoplast fusion, complementary frequency between *B. brongniartii* Type 2 and *B. amorphi* was high, but those of other pairing combinations were low.

V. CONCLUSION

Cell fusion procedures provide a powerful and efficient means for obtaining new types of invertebrate cell cultures. Moreover, the mechanisms are now possible where the components of one cell of the hybrid can be intermixed with those of its fused partner in a natural manner: genetic exchange, virus uptake, and the effect of drugs can be passed from one cell to the other. The state of the art of cell fusion techniques has progressed to the point where physical fusion procedures provide control and reproducibility never before realized. This progression may eliminate the need for exogenous biological and chemical fusion-promoting agents. Applications for invertebrate cell systems using fusion procedures appear to be limited only by the creativity of the research scientist.

REFERENCES

1. **Okada, Y., Suzuki, T., and Husaka, Y.**, Interaction between influenza virus and Ehrlich's tumor cells. II. Fusion phenomenon of Ehrlich's tumor cells by the action of HVJ Z strain, *Med. J. Osaka Univ.*, 7, 709, 1957.
2. **Ishida, N. and Homma, M.**, Sendai Virus, *Adv. Virus Res.*, 23, 349, 1978.
3. **Barski, G., Sorieul, S., and Confert, F.**, Production dans des cultures in vitro de deux souches cellulaires en association de cellules de caractère "hybride", *C. R. Acad. Sci.*, 251, 1825, 1960.
4. **Sorieul, S. and Ephrussi, B.**, Karyological demonstration of hybridization of mammalian cells in vitro, *Nature (London)*, 190, 653, 1961.
5. **Harris, H.**, *Cell Fusion, The Dunham Lectures*, Oxford University Press, London, 1970.
6. **Harris, H. and Watkins, J. F.**, Hybrid cells derived from mouse and man: Artificial heterokaryons of mammalian cells from different species, *Nature (London)*, 205, 640, 1965.
7. **Ringertz, N. R. and Savage, R. E.**, *Cell Hybrids*, Academic Press, New York, 1976.
8. **Köhler, G. and Milstein, C.**, Continuous cultures of fused cells secreting antibody of predefined specificity, *Nature (London)*, 256, 495, 1975.
9. **Ahkong, Q. F., Howell, J. I., Lucy, J. A., Safwat, F., Davy, M. R., and Cocking, E. C.**, Fusion of hen erythrocytes with yeast protoplasts induced by polyethylene glycol, *Nature (London)*, 255, 66, 1975.
10. **Cocking, E. C.**, Protoplasts: past and present, in *Advances in Protoplast Research*, Ferecnzy, L. and Farkas, G. L., Eds., Pergamon Press, Oxford, 1980, 3.
11. **Cocking, E. C.**, Plant-animal cell fusions, in *Cell Fusion*, Ciba Found. Symp. 103, Evered, D. and Whelan, J., Eds., Pitman, London, 1984, 119.
12. **Zimmerman, U., Vienkin, J., Pilwat, C., and Arnold, W. M.**, Electro-fusion of cells: Principles and potential for the future, in *Cell Fusion*, Ciba Found. Symp. 103, Evered, D. and Whelan, J., Eds., Pitman, London, 1984, 60.
13. **Neuman, E. and Bierth, P.**, Gene transfer by electroporation, *Am. Biotechnol. Lab.*, March/April, 10, 1986.
14. **Schierenberg, E.**, Laser-induced cell fusion, in *Cell Fusion*, Sowers, A. E., Ed., Plenum Press, New York, 1987, 409.
15. **Weber, H., Förster, W., Jacob, H.-E., and Berg, H.**, Microbiological implications of electric field effects. III. Stimulation of yeast protoplast fusion by electric field pulses, *Z. Allg. Mikrobiol.*, 21, 555, 1981.
16. **Weber, H., Förster, W., Berg, H., and Jacob, H.-E.**, Parasexual hybridization of yeasts by electric field stimulated fusion of protoplasts, *Curr. Genet.*, 4, 165, 1981.
17. **Auer, D., Brandner, G., and Bodemer, W.**, Dielectric breakdown of the red blood cell membrane and uptake of SV 40 DNA and mammalian RNA, *Naturwissenschaften*, 63, 391, 1976.
18. **Zimmerman, U., Riemann, F., and Pilwat, G.**, Enzyme loading of electrically homogeneous human red blood cell ghosts prepared by dielectric breakdown, *Biochim. Biophys. Acta*, 436, 460, 1976.
19. **Barigozzi, C.**, I. *Drosophila* cells in vitro: behavior and utilization for genetic purpose, *Curr. Top. Microbiol. Immunol.*, 55, 209, 1971.
20. **Nakajima, S. and Miyake, T.**, Cell fusion between temperature-sensitive mutants of a *Drosophila melanogaster* cell line, *Somatic Cell Genet.*, 4, 131, 1978.
21. **Becker, J.-L.**, Fusions in vitro de cellules somatiques en culture de *Drosophila melanogaster*, induites par la concanavaline A, *C. R. Acad. Sci., (Paris)*, 275, 2969, 1972.

22. **Bernhard, H. P.**, *Drosophila* cells: Fusion of somatic cells by polyethylene glycol, *Experientia*, 32(Abstr.), 786, 1976.

23. **Halfer, C. and Petrella, L.**, Cell fusion induced by lysolecithin and concanavalin A in *Drosophila melanogaster* somatic cells cultured in vitro, *Exp. Cell Res.*, 100, 399, 1976.

24. **Rizki, R. M., Rizki, T. M., and Andrews, C. A.**, *Drosophila* cell fusion induced by wheat germ agglutinin, *J. Cell Sci.*, 18, 113, 1975.

25. **Wyss, C.**, TAM selection of *Drosophila* somatic cell hybrids, *Somatic Cell Genet.*, 5, 29, 1979.

26. **Mitsuhashi, J.**, Fusion of insect cells, in *Biotechnology Advances in Invertebrate Pathology and Cell Culture*, Maramorosch, K., Ed., Academic Press, New York, in press.

27. **Wyss, C.**, Cell hybrid analysis of ecdysone sensitivity and resistance in Drosophila cell lines, in *Invertebrate Systems in Vitro*, Kurstak, E., Maramorosch, K., and Dübendorfer, A., Eds., Elsevier/North-Holland, New York, 1980, 279.

28. **Tsang, K. R. and Brooks, M. A.**, Neoplasias caused by cultured insect cells, in *Invertebrate Systems in Vitro*, Kurstak, E., Maramorosch, K., and Dübendorfer, A., Eds., Elsevier/North-Holland, New York, 1980, 535.

29. **Johnson, R. T., Rao, P. N., and Hughes, H. D.**, Mammalian cell fusion. III. A HeLa cell inducer of premature chromosome condensation active in cells from a variety of animal species, *J. Cell. Physiol.*, 76, 151, 1970.

30. **Conover, J. H., Zepp, H. D., Hirschhorn, K., and Hodes, H. L.**, Production of human-mosquito somatic cell hybrids and their response to virus infection, *Curr. Top. Microbiol. Immunol.*, 55, 85, 1971.

31. **Mazzone, H. M., Breillatt, J. P., and Anderson, N. G.**, Zonal rotor purification of a nuclear polyhedrosis virus of the European pine sawfly (*Neodiprion sertifer*, Geoffroy), in *Proc. 4th Int. Colloq. Insect Pathol.*, College Park, MD, 1970, 371.

32. **Podgwaite, J. D.**, Strategies for field use of Baculoviruses, in *Viral Insecticides for Biological Control*, Maramorosch, K. and Sherman, K. E., Eds., Academic Press, New York, 1985, 775.

33. **Weiss, M. C. and Green, H.**, Human-mouse hybrid cell lines containing partial complements of human chromosomes and functioning human genes, *Proc. Natl. Acad. Sci. U.S.A.*, 58, 1104, 1967.

34. **Pontecorvo, G.**, Induction of directional chromosome elimination in somatic cell hybrids, *Nature (London)*, 230, 367, 1971.

35. **Podgwaite, J. D. and Mazzone, H. M.**, Latency of insect viruses, *Adv. Virus Res.*, 31, 293, 1986.

36. **Mazzone, H. M.**, Cultivation of gypsy moth hemocytes, *Curr. Top. Microbiol. Immunol.*, 55, 198, 1971.

37. **Mazzone, H. M.**, Influence of polyphenol oxidase on hemocyte cultures of the Gypsy moth, in *Invertebrate Tissue Culture*, Kurstak, E. and Maramorosch, M., Eds., Academic Press, New York, 1976, 275.

38. **Shivarova, N. and Grigorova, R.**, Microbiological implications of electric field effects. VIII. Fusion of *Bacillus thuringiensis* protoplasts by high electric field pulses, *Bioelectrochem. Bioenerg.*, 11, 181, 1983.

39. **Kuta, A. E., Rhine, R. S., and Heebner, G. M.**, Electrofusion. A new tool for Biotechnology, *Am. Biotechnol. Lab.*, May/June, 31, 1985.

40. **Shimizu, S.**, Protoplast fusion of insect pathogenic fungi, in *Biotechnology Advances in Invertebrate Pathology and Tissue Culture*, Maramorosch, K., Ed., Academic Press, New York, in press.

41. **Zimmerman, U. and Greyson, J.**, Electric field-induced cell fusion, *Biotechniques*, 1, 118, 1983.

42. **Zimmerman, U. and Vienken, J.**, Electric field-induced cell-to-cell fusion, *J. Membr. Biol.*, 67, 165, 1982.

43. **Poste, G. and Nicholson, G. L.**, Membrane Fusion, in *Cell Surface Reviews*, Vol. 5, Elsevier/North-Holland, Amsterdam, 1978.

44. **Zimmerman, U. and Steudle, E.**, Physical aspects of water relations of plant cells, *Adv. Bot. Res.*, 6, 45, 1978.

45. **Power, J. B., Evans, P. K., and Cocking, E. C.**, Fusion of plant protoplasts, in *Membrane Fusion*, Poste, G. and Nicholson, G. L., Eds., Elsevier/North-Holland, Amsterdam, 1978, 369.

Chapter 16

GENE ANALYSIS BY BLOT HYBRIDIZATION ON THE SILKWORM, *BOMBYX MORI*, CELL LINES

O. Ninaki, N. Takada, H. Fujiwara, T. Ogura, N. Miyajima, A. Kiyota, and H. Maekawa

TABLE OF CONTENTS

I. INTRODUCTION

In the silkworm, *Bombyx mori*, three cell lines (S.P.C.Bm36,[1] BM-N,[2] and SES-BoMo-15A[3]) have been established. These cell lines were already characterized, but genetic and physiologic data are rare. Furthermore, one of them, S.P.C.Bm36(Bm36) cell line in the currently available form, is not *B. mori* type.[4,5] The original cell source of the BM-N(Bm) has not been determined. Thus, we tried to establish and characterize new cell lines of *B. mori*. An appropriate method to characterize the genome of the new cell lines and the previously established cell lines is blot hybridization with the gene probes proper to *B. mori*. It is a useful method for biochemical analysis.

II. MATERIALS AND METHODS

DNA was extracted from the cultured cells of several cell lines by the methods of Suzuki et al.[6] and from the posterior silk glands of the last instar larvae of *Bombyx* strains and *Samia cynthia ricini* by use of the "Freezer-Mill"(Spex, U.S.).[7] DNA from *Drosophila melanogaster* flies, liver of *Xenopus laevis*, and a human cell line were gifts of Drs. K. Hiromi, A. Saito, and R. Kominami, respectively.

DNA was digested with restriction endonucleases (Takara Shuzo Co., Ltd., Kyoto) and transferred to nitrocellulose filters, BA85 (Schleicher & Schuell, NH), by the Southern blotting method.[8] Hybridization was carried out in a solution containing $0.6\,M$ NaCl, $8\,mM$ EDTA, 120 mM Tris-HCl (pH 8.0), and 0.2% SDS at 10 × Denhardt[9] 65°C for 16 h or more with DNA probes. Ribosomal DNA (rDNA) clones, pBmR145 and -161, are subclones from λBmR11.[10] The pBmF6 clone is derived from pFb29 of fibroin gene clone.[11] The *Bombyx* multicopy gene (BMC) clone, pSB2A, is a subclone from BMC-1.[12] DNA probes were prepared by nick translation of cloned DNA with ^{32}P-dCTP (Amersham, U.K.). An autoradiograph was obtained by exposing Kodak® X-Omat XAR film to the hybridized filters for 1 or 2 d.

The 5′-end sequence of 5.8S rRNA was determined by Donis-Keller's enzymatic analysis.[13]

III. RESULTS

A. Establishment of New Cell Lines

Two cell lines of SES-Bm-e 21 and SES-Bm-1 30 were established from the strain e 21(*pe, re, ch, Sph*) and 1 30(*pM, Ze, L, q*) of *B. mori*, respectively. Dissected embryos (ten embryos, at the stage of 1 d before hatching, per dish) of each strain were first cultured in IPL-41 medium[14,15] containing 15% fetal bovine serum (FBS) for 2 weeks at 25°C in the dark, and the tissues were cultured for 3 months in 10% FBS. One year was required for the satisfactory proliferations of these cells. SES-Bm-e 21 has currently proliferated with the doubling time of 2 to 3 d, and SES-Bm-1 30 has grown only in the aggregated form with doubling time of 3 to 4 d.

B. Characterization of ribosomal DNA (rDNA)

DNA extracted from each source was digested with several restriction enzymes and blotted to the nitrocellulose filter after electrophoresis. As the autoradiograph of the hybridization with the rDNA probe shows in Figure 1, similar patterns were observed between each pair of the Bm cell line and the P-22 strain: the e 21 cell and the strain of Gunpo × Shugyoku, and the 1 30 cell line and *B. mandarina*, respectively. Major bands of them are identical to one another among the *Bombyx* DNAs described above, while the minor bands reveal slight differences among them. These results might be caused by the polymorphism of the multigene family. In *B. mandarina*, however, additional major bands were observed.[5]

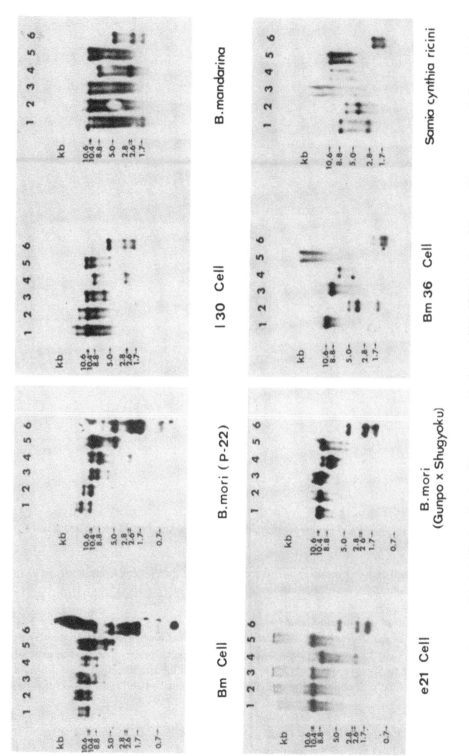

FIGURE 1. Autoradiograph of blot hybridization of various DNA sources with ribosomal DNA (rDNA) probes of *B. mori*. The source name is illustrated below each photograph.

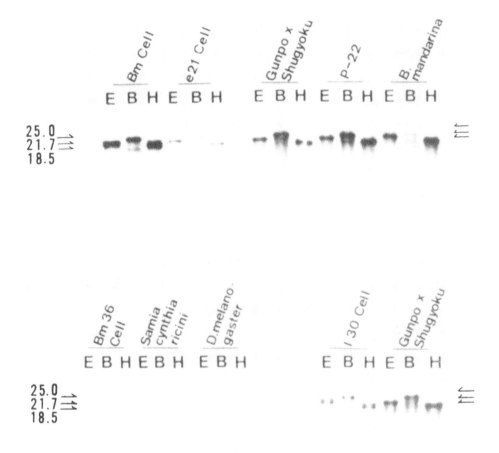

FIGURE 2.   Autoradiograph of blot hybridization of various DNA sources with the fibroin gene probe of *B. mori.*

In the Bm36 cell and *S. cynthia ricini*, there are distinct differences from the *B. mori* in the major bands of rDNA derived from the cell lines and silk glands (Figure 1).

C. Characterization of Fibroin Gene

The detection of the fibroin gene was carried out in the extracted DNA the same way as the case of rDNA, by blot hybridization using the fibroin gene probe. The fibroin gene of *B. mori* is unique to *Bombyx* species but was not detected in the Bm36 cell, *S. cynthia ricini* and *D. melanogaster* as shown in Figure 2. The newly established cell lines did contain the fibroin gene.

D. Detection of *Bombyx* Multicopy Gene (BMC) Family

The BMC family is one of the multigene families[12] and is also unique to the *Bombyx* species as the results of the blot hybridization with the probe of the BMC gene show (Figure 3). The homologous sequences to the family are not detected in other species of Lepidoptera, Diptera, and vertebrates indicated in Figure 3. This unique sequence was also not included in the Bm36 cell line.

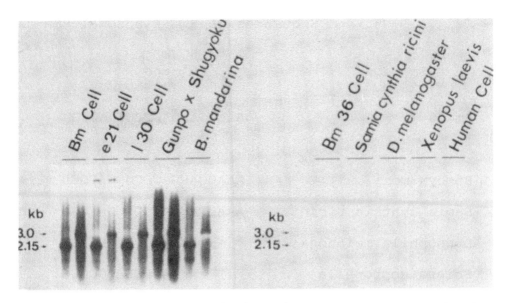

FIGURE 3.   Autoradiograph of blot hybridization of various DNA sources with the BMC gene probe of the *B. mori* BMC-1 family.

E. Sequence Homology of 5.8S rRNA

5.8S rRNA sequences are conserved among the overall species, but differences of the 5'-end have been reported as shown in Figure 4. The sequence of the Bm36 cell line is identical to that of *B. mori*[13] and very similar to *S. cynthia ricini*; however, it is distinct from that of *D. melanogaster* and *Sciara coprophila*.[16]

IV. DISCUSSION

We used the blot hybridization method for the characterization of established cell lines. This method is very useful and secure, in addition to the biochemical analysis, such as the isozyme assay.

B. mandarina may be one of the ancestors of *B. mori*, and it is possible to mate them. In the detection of rDNA, fibroin gene, and BMC family, there are few differences between both species. Thus, this method is able to detect differences among the related species. The patterns of rDNA and fibroin gene of the new cell lines (Figure 2) were apparently changed because the selection from the mixed population of the original strains with the genetic polymorphism[17] was caused through the establishment. This evidence reveals that the slight differences at gene level among the strains is distinguishable by the blot hybridization method.

From the results of the gene detection, we concluded that the currently available Bm36 cell line is not derived from *Bombyx* but might be a species closely related to *Bombyx* because of the sequence data of the 5.8S rRNA. One possibility has been proposed that this cell line may be derived from *Antheraea* judging from the fibroin gene assay.[18]

We hope investigators will use the two new cell lines of *B. mori* in experiments at the level of molecular biology.

ACKNOWLEDGMENTS

We are indebted to the following people who supplied the DNA and cells for our use: Drs. K. Aizawa, E. Ishikawa, Y. Suzuki, K. Hiromi, R. Kominami, J. Kusuda, A. Saito, and T. Shimada, and Mr. K. Mikitani and M. Sudo. This study was partly supported by grants from the Ministry of Agriculture, Forestry and Fisheries; the Ministry of Education,

5'-end Sequences of 5.8S Ribosomal RNA

Species	5'-end
Bm 36	AAAUGAUUACCCUGGACGGU
Bombyx mori	AAAUGAUUACCCUGGACGGU
Samia cynthia ricini	AAACCAUUACCCUGGACGGU
Drosophila melanogaster	AACUCUAAGCGGU
Sciara coprophila	AACCCUAAGCGGG
Acyrthosiphon magnoliae	CGUCCCCGAACGGC
Xenopus laevis	UCGACUCUUAGCGGU
Saccharomyces cerevisiae	AAACUUUCAACAAC

FIGURE 4.   5'-End sequences of 5.8S ribosomal RNA (rRNA) of various species.

Science and Culture of Japan, and the Foundation for Promotion of Cancer Research supported by the Japan Ship Building Industry Foundation.

REFERENCES

1. **Quiot, J. M.,** Establishment of a cell line (S.P.C. Bm36) from the ovaries of *Bombyx mori* L.(Lepidoptera), *Sericologia,* 22, 25, 1982.
2. **Volkman, L. E. and Goldsmith, P. A.,** Generalized immunoassay for *Autographa californica* nuclear polyhedrosis virus infectivity *in vitro, Appl. Environ. Microbiol.,* 44, 227, 1982.
3. **Inoue, H. and Mitsuhashi, J.,** A *Bombyx mori* cell line susceptible to a nuclear polyhedrosis virus, *J. Seric. Sci. Jpn.,* 53, 108, 1984.
4. **Suzuki, Y.,** personal communication, 1987.
5. **Maekawa H., Takada, N., Mikitani, K., Ogura, T., Miyajima, N., Fujiwara, H., Kobayashi, M., and Ninaki, O.,** Nucleolus organizers in the wild silkworm *Bombyx mandarina* and the domesticated silkworm *B. mori, Chromosoma (Berlin),* 96, 263, 1988.
6. **Suzuki, Y., Gage, L. P., and Brown, D. D.,** The genes for silk fibroin in *Bombyx mori,* J. Mol. Biol., 70, 637, 1972.
7. **Takada, N., Miyajima, N., Ogura, T., and Maekawa, H.,** The high molecular weight DNA prepared from the posterior silk gland of *Bombyx mori* by use of "Freezer-Mill" (in Japanese with English summary), *Jpn. J. Appl. Entomol. Zool.,* 29, 83, 1985.
8. **Southern, E. M.,** Detection of specific sequences among DNA fragments separated by gel electrophoresis, *J. Mol. Biol.,* 98, 502, 1975.
9. **Denhardt, D. T.,** A membrane-filter technique for the detection of complementary DNA, *Biochem. Biophys. Res. Commun.,* 23, 641, 1966.

10. **Fujiwara, H., Ogura, T., Takada, N., Miyajima, N., Ishikawa, H., and Maekawa, H.,** Introns and their franking sequences of *B. mori* rDNA, *Nucleic Acids Res.,* 12, 6861, 1984.
11. **Tsujimoto, Y. and Suzuki, Y.,** Structural analysis of the fibroin gene at the 5' end and its surrounding regions, *Cell,* 16, 425, 1979.
12. **Ogura, T.,** unpublished data, 1987.
13. **Donis-Keller, H.,** Phy M: an RNase activity specific for U and A residues useful in RNA sequence analysis, *Nucleic Acids Res.,* 8, 3133, 1980.
14. **Dougherty, E. M., Weiner, R. M., Vaughn, J. T., and Reichelderfer, C. F.,** Physical factors that affect *in vitro Autographa californica* nuclear polyhedrosis virus infection, *Appl. Environ. Microbiol.,* 41, 1166, 1981.
15. **Goodwin, R. H.,** personal communication, 1984.
16. **Erdmann, V. A., Woltecs, J., Huysmans, E., and Wachter, R. D.,** Collection of published 5S, 5.8S and 4.5S ribosomal RNA sequences, *Nucleic Acids Res.,* 14, r59, 1986.
17. **Sprague, K. U., Roth, B. M., Manning, R. F., and Gage, L. P.,** Alleles of the fibroin gene coding for proteins of different lengths, *Cell,* 17, 407, 1979.
18. **Tamura, T.,** personal communication, 1987.

Chapter 17

GENETICS OF MOSQUITO CELLS: CLONAL ANALYSIS AND USE OF CLONING VECTORS IN INSECT CELLS

Ann Marie Fallon

TABLE OF CONTENTS

I. INTRODUCTION

The contribution of insect systems to developmental biology and genetics is best exemplified by the vast amount of information that has been accumulated using the fruit fly, *Drosophila melanogaster*. Biological features of this organism that have facilitated laboratory investigation include its short life cycle, simple dietary requirements, polytene salivary gland chromosomes that are readily amenable to cytological procedures, and a small genome size. These properties have contributed to the isolation and characterization of a vast repertoire of mutant flies, initially using classical genetic approaches. More recently, the development of techniques for transforming *Drosophila* cells in culture, and for transforming *Drosophila* embryos using vectors based on the transposable P element, has provided powerful new approaches to carrying out fundamental molecular investigations using this insect. Clearly, the application of the techniques and approaches that have been successful with *Drosophila* to insect species of medical or agricultural importance constitutes an important challenge to insect physiologists and biochemists.

Our own work focuses on the mosquito, a disease vector whose impact on world health has increased significantly in recent years due to the appearance of insecticide-resistant strains.[1] Our approach has been to use cultured cells as an important adjunct to physiological studies at the level of the intact organism. A brief survey of recent developments in our laboratory is described below; more comprehensive reviews of genetic and biochemical studies that have been carried out using mosquito cells in culture include those of Kurtti and Munderloh[2] and Fallon and Stollar.[3]

II. DESCRIPTION OF THE *AEDES ALBOPICTUS* CELLS

The *Aedes albopictus* cells used in our laboratory were derived from minced neonate larvae[4] and were initially maintained in the insect tissue culture medium described by Mitsuhashi and Maramorosch.[5] Sarver and Stollar[6] have more recently adapted these cells to Eagle's[7] vertebrate tissue culture medium. For routine use, we maintain *A. albopictus* cells in Eagle's minimal medium supplemented with nonessential amino acids, glutamine, penicillin, streptomycin, and heat-inactivated fetal calf serum at a final concentration of 5%. Cells are maintained at 28°C in a 5% CO_2 atmosphere. Under these conditions, standard ("wild type") cells have a doubling time of approximately 18 h.[3]

III. MUTANT CELL LINES

Stollar and Mento were the first to apply systematically the techniques and approaches of somatic cell genetics to mosquito cells in culture.[8] Since the initial description of mutants resistant to bromodeoxyuridine (thymidine-kinase deficient), a number of mutants affecting various intracellular processes have been accumulated, including cells resistant to α-amanitin,[8] ouabain,[8] methotrexate,[9] cycloheximide,[10] and puromycin.[11] In the case of the antibiotics cycloheximide and puromycin, which act at the level of the ribosome, it has not yet been possible to obtain from vertebrate cells in culture mutants with comparable levels of resistance. Resistance to these drugs can be demonstrated in cell-free lysates prepared from the mutant mosquito cells, indicating that the alteration in these mutants affects an intracellular component of the protein synthetic machinery, and not the permeability properties of the plasma membrane.[10,11]

A long-term goal of our research is the isolation and characterization of mosquito ribosomal protein genes. Cloned genes can be used as probes to investigate physiological processes in the intact organism. Specifically, in the mosquito, the blood meal triggers a cascade of hormonal events that culminate in the synthesis of egg yolk protein (vitellogenin) by the fat body.[12] Vitellogenin is released from the fat body into the hemolymph and is selectively sequestered by the developing oocytes. Early work showed that this reproductive process included a cycle of accumulation and degradation of total RNA in the fat body of blood-fed mosquitoes.[13] More recently, we have shown that the amount of total fat body RNA represents synthesis and degradation of ribosomes,[14] which presumably enable the fat body to synthesize massive amounts of vitellogenin. Recent analysis of ultrastructural changes in mosquito fat body during the vitellogenic cycle were consistent with these biochemical data.[15]

One approach to cloning ribosomal protein genes has been to examine antibiotic-resistant mutants, such as those resistant to cycloheximide or to puromycin, for a ribosomal protein with altered electrophoretic mobility. Using the two-dimensional polyacrylamide gel system described by Sanders et al.,[16] we have identified 28 to 31 ribosomal proteins from the small subunit, and 36 to 39 proteins from the large ribosomal subunit isolated from *A. albopictus* cells.[17] The mosquito ribosomal proteins range in mass from 10 to 53 kDa. In cultured cells, the largest protein on the small ribosomal subunit, S1, is the major phosphoribosomal protein. *Aedes* S1 appears to share homology at the level of peptide mapping with S6, the major phosphoribosomal protein in *D. melanogaster* cells.[17] The ribosomal proteins from *A. aegypti* fat body are virtually identical in electrophoretic mobility to those from *A. albopictus* cells in culture. However, in fat body, two proteins (S10 and A) in addition to S1 are phosphorylated.[14]

The two-dimensional gel system has also been used to compare the ribosomal proteins from cycloheximide-resistant and sensitive mosquito cells. Although we have not been able to identify an altered protein from resistant cells, our most recent studies have shown that the cycloheximide-resistant cells are temperature sensitive for growth at 34.5°C and have increased sensitivity to G-418, another antibiotic that acts at the level of the ribosome.[18] Work is in progress to obtain additional clones of *A. albopictus* cells with increased resistance to cycloheximide. In similar studies with emetine-resistant Chinese hamster cells, second-level mutants with increased resistance to the selective agent (relative to the initial mutants) facilitated identification of the altered ribosomal protein.[19]

IV. MOLECULAR ANALYSIS OF MOSQUITO GENES

Our second approach to identifying ribosomal protein genes has been to clone into a plasmid vector, double-stranded cDNA copies of size-selected mRNA from *Aedes* cells. Using the technique of hybrid selection, followed by *in vitro* translation of the selected mRNA in the wheat germ cell-free translation system, we have recently identified cDNA sequences corresponding to ribosomal proteins L8, L14, and L31.[23] Using these cDNA probes, ribosomal protein gene expression will be investigated in fat body. In addition, the cDNAs will be used to identify the corresponding gene from a bacteriophage λ library.

To develop the technology for studying expression of cloned genes in *Aedes* cells, we have attempted to introduce into these cells various recombinant vectors using the calcium phosphate coprecipitate method, which has been widely used with vertebrate cells. We found that transfection of mosquito cells was more effective when the poly-cation polybrene (1,5-dimethyl-1,5-diazaundecamethylene polymethobromide) rather than calcium phosphate was used as the chemical mediator.[20,21] The DNA construct used to establish our transfection protocol contained a heat-inducible promoter from the *Drosophila* heat shock protein 70 gene, fused in frame with the chloramphenicol acetyltransferase gene from *Escherichia coli*.[22] Since the heat shock proteins show considerable conservation during evolution, it was not surprising that the *Drosophila* heat shock protein promoter functions in mosquito cells. Interestingly, however, we have not yet detected expression in transfected mosquito cells of genes under the control of the promoter from the *Drosophila* transposable element *copia*.

Development of species-specific vectors, and establishment of techniques for stable integration of transfected DNA into insect cells other than those of *Drosophila*, requires not only the establishment and optimization of transfection protocols, but also the identification of effective promoters to drive expression of introduced genes. It is our hope that our efforts towards identifying regulatory sequences from mosquito ribosomal protein genes will ultimately contribute to the development of new genetic tools for investigation of mosquito biology.

REFERENCES

1. **Mouchès, C., Pasteur, N., Bergé, J. B., Hyrien, O., Raymond, M., de St. Vincent, B. R., de Silvestri, M., and Georghiou, G. P.,** Amplification of an esterase gene is responsible for insecticide resistance in a California *Culex* mosquito, *Science*, 233, 778, 1986.
2. **Kurtti, T. J. and Munderloh, U. G.,** Mosquito cell culture, in *Advances in Cell Culture*, Vol. 3, Maramorosch, K., Ed., Academic Press, New York, 1984, 259.
3. **Fallon, A. M. and Stollar, V.,** The biochemistry and genetics of mosquito cells in culture, in *Advances in Cell Culture*, Vol. 5, Maramorosch, K., Ed., Academic Press, New York, 1987, 97.
4. **Singh, K. R. P.,** Cell cultures derived from larvae of *Aedes albopictus* (Skuse) and *Aedes aegypti* (L.), *Curr. Sci.*, 36, 506, 1967.
5. **Mitsuhashi, J. and Maramorosch, K.,** Leafhopper tissue culture: embryonic nymphal and imaginal tissues from aseptic insects, *Contrib. Boyce Thompson Inst.*, 22, 435, 1964.
6. **Sarver, N. and Stollar, V.,** Sindbis virus-induced cytopathic effect in clones of *Aedes albopictus* (Singh) cells, *Virology*, 80, 390, 1977.
7. **Eagle, H.,** Amino acid metabolism in mammalian cell cultures, *Science*, 130, 432, 1959.
8. **Mento, S. J. and Stollar, V.,** Isolation and partial characterization of drug-resistant *Aedes albopictus* cells, *Somatic Cell Genet.*, 4, 179, 1978.
9. **Fallon, A. M.,** Methotrexate resistance in cultured mosquito cells, *Insect Biochem.*, 14, 697, 1984.
10. **Fallon, A. M. and Stollar, V.,** Isolation and characterization of cycloheximide-resistant mosquito cell clones, *Mutat. Res.*, 96, 201, 1982.
11. **Fallon, A. M. and Stollar, V.,** Isolation and characterization of puromycin-resistant clones from cultured mosquito cells, *Somatic Cell Genet.*, 8, 521, 1982.
12. **Racioppi, J. V., Gemmill, R. M., Kogan, P. H., Calvo, J. M., and Hagedorn, H. H.,** Expression and regulation of vitellogenin messenger RNA in the mosquito, *Aedes aegypti*, *Insect Biochem.*, 16, 255, 1986.
13. **Hagedorn, H. H., Fallon, A. M., and Laufer, H.,** Vitellogenin synthesis by the fat body of the mosquito *Aedes aegypti*: evidence for transcriptional control, *Dev. Biol.*, 31, 281, 1973.
14. **Hotchkin, P. G. and Fallon, A. M.,** Ribosome metabolism during the vitellogenic cycle of the mosquito, *Aedes aegypti*, *Biochim. Biophys. Acta*, 924, 352, 1987.
15. **Raikhel, A. S. and Lea, A. O.,** Pre-vitellogenic development and vitellogenin synthesis in the fat body of a mosquito: an ultrastructural and immunocytochemical study, *Tissue Cell*, 15, 281, 1983.
16. **Sanders, M. M., Groppi, V. E., Jr., and Browning, E. T.,** Resolution of basic cellular proteins including histone variants by two-dimensional gel electrophoresis: evaluation of lysine to arginine ratios and phosphorylation, *Anal. Biochem.*, 103, 157, 1980.
17. **Johnston, A. M. and Fallon, A. M.,** Characterization of the ribosomal proteins from mosquito (*Aedes albopictus*) cells, *Eur. J. Biochem.*, 150, 507, 1985.
18. **Nouri, N. and Fallon, A. M.,** Pleiotropic changes in cycloheximide-resistant insect cell clones, *In Vitro Cell. Dev. Biol.*, 23, 175, 1987.
19. **Madjar, J.-J., Frahm, M., McGill, S., and Roufa, D.,** Ribosomal protein S14 is altered by two-step emetine resistance mutations in Chinese hamster cells, *Mol. Cell. Biol.*, 3, 190, 1983.
20. **Durbin, J. E. and Fallon, A. M.,** Transient expression of the chloramphenicol acetyltransferase gene in cultured mosquito cells, *Gene*, 36, 173, 1985.
21. **Fallon, A. M.,** Factors affecting polybrene-mediated transfection of cultured *Aedes albopictus* (mosquito) cells, *Exp. Cell Res.*, 166, 535, 1986.
22. **Di Nocera, P. P. and Dawid, I. B.,** Transient expression of genes introduced into cultured cells of *Drosophila*, *Proc. Natl. Acad. Sci. U.S.A.*, 80, 7095, 1983.
23. **Durbin, J. E., Swerdel, M. R., and Fallon, A. M.,** Identification of cDNAs corresponding to mosquito ribosomal protein genes, *Biochim. Biophys. Acta*, 950, 182, 1988.

Chapter 18

HEAT SHOCK- AND ECDYSTERONE-INDUCED PROTEIN SYNTHESIS IN *DROSOPHILA* CELLS

Edward M. Berger and Karen M. Rudolph

TABLE OF CONTENTS

I. INTRODUCTION

Drosophila melanogaster is one of the best described eukaryotes, in terms of its genetics and cytogenetics. However, because of small size, its usefulness for studies at the molecular level has been somewhat restricted. This is especially true in the area of endocrinology, in particular, the molecular genetic events underlying metamorphosis.

It has been known for some time that insect development is regulated by the interaction of two hormones: juvenile hormone, a sesquiterpene; and ecdysterone (also known as 20-hydroxyecdysone or β-ecdysone), a steroid. When both hormones are present during a molt, development proceeds from either larva to larva, or larva to pupa. A pupal molt occurring in the presence of ecdysterone alone leads to the development of adult structures. The discovery of polytene chromosomes in dipteran larval tissues enabled investigators to clearly establish that underlying the developmental transformations were distinct changes in the pattern of gene expression. These changes could actually be visualized in the microscope, as the sequential appearance and regression of so-called chromosome puffs; i.e., localized regions of active RNA transcription. It was soon shown that the addition of ecdysterone to medium containing salivary glands taken from young third instar larvae led to a stereotyped pattern of puff formation and regression.[1] As expected, these patterns faithfully reproduced the ontogenetic pattern seen *in vivo*. Later studies found that other organs could be cultured *in vitro* and that each responded to hormone in a tissue-specific manner.

From these studies emerged a general model for ecdysterone action at the molecular genetic level. It involved the association of free hormone with a receptor in the cell and the subsequent association of the hormone-receptor complex with target genes in the nucleus. Binding to chromatin, in a DNA sequence-specific way, somehow led to gene activation or repression. Because of size limitations, the appropriate molecular studies needed to confirm aspects of the model could not be carried out. Meanwhile, mammalian endocrinologists continued to exploit easily manipulable cell and organ culture systems, and proceeded to unravel the detailed mechanism of steroid hormone action using estrogen-, progesterone-, and glucocorticoid-responsive systems.   Revival of interest in *Drosophila* as a model system to study ecdysterone action came in 1969 when two laboratories independently established the first series of continuous cell lines derived from embryonic tissue.[2,3] These cell lines have become critically important tools for the application of modern molecular techniques to the question of ecdysterone action. This article will summarize that progress, focusing, in the end, on an unexpected observation — that among the ecdysterone-inducible genes identified using cell lines are a linked cluster that can be independently activated by heat shock or other stress conditions.

II. ECDYSTERONE RESPONSES

A. Morphological Changes

Morphological transformations are a characteristic ecdysterone response of nearly every unselected cell line examined.[4-8] The cells, which are roughly spherical and loosely attached to their substrate, become tightly adherent and flattened. Within 4 h, the cells begin to extend and retract a complex array of filipodia that participate in the onset of cell motility. Motile cells soon begin migrating into foci, forming large, adherent aggregates that become distributed evenly along the substrate. Each aggregate contains hundreds to thousands of cells. The entire process, when visualized in time lapse, resembles *Dictyostelium* aggregation. Scanning EM studies have shown that aggregated cells are covered by an amorphous matrix material of unknown composition.[5]

A second characteristic response of all unselected cell lines is the arrest of cell division.[4-8,11,12] Apparently, cells complete one round of DNA replication and become arrested in G2. DNA synthesis and histone mRNA content decrease dramatically during the first day.[9,13] Cherbas et al. have shown that the commitment to mitotic arrest occurs within 15 min of exposure and does not require the further presence of hormone in the medium. Ecdysteroid analogues with little or no biological activity do not produce this arrest.[10]

There is evidence[14,15] that in certain cell lines mitotic activity resumes after 5 or 6 d in hormone and is accompanied by a period in which cells become refractory to ecdysterone. This insensitivity is based on the loss of ecdysterone-receptor activity and is reversed after several weeks of culture without ecdysterone. Other cell lines do not seem to enter this refractory phase. From these lines one can directly select proliferating clonal cell lines, in soft agar, that are insensitive to ecdysterone.[16,17] This insensitivity, while also based on the absence of ecdysteroid receptor activity,[18,19] is not reversed. These ecdysterone-insensitive lines have become useful tools for discriminating between *bona fide* hormone-dependent effects and effects of a nonspecific nature.

B. Molecular Correlates

The rapid and dramatic morphological transformation of *Drosophila* cells, in response to ecdysterone, led us to suspect accompanying and underlying changes at the molecular level produced by hormone treatment. Prime suspects were (1) the cytoplasmic actins, major components of the microfilament system; (2) the tubulins, structural components of microtubules; and (3) cell surface glycoproteins that would include specific aggregation factors.

1. Actin Isoforms

The abrupt onset of cell motility in hormone-treated cells suggested the possibility of microfilament production or reorganization. To test this, we and others[20-24] began to look into the types and amounts of cytoplasmic actins present before and after treatment. Initial studies focused on the identification of *Drosophila* actins in labeling experiments using two-dimensional polyacrylamide gel electrophoresis (2-D PAGE). We found that there were two isoforms and established that one was a metabolic precursor of the other.[25] We later showed that the posttranslational modification involved actin acetylation.[26] We next measured actin content, synthesis, and degradation in S3 cells before, during, and after ecdysterone treatment to determine whether the increased motility was due to an increase or other change in actin content. The cytoplasmic actin content of responsive cells was found to double during hormone-induced cell elongation. The elevated level of actin is due to a ninefold increase in the level of cytoplasmic actin gene A3 mRNA and, at most, a twofold increase in cytoplasmic actin gene A2 mRNA. Muscle-specific actin genes A1, A4, A5, and A6 are not expressed in these cells.

2. Tubulin Isoforms

The elongation response suggested that ecdysterone might produce some qualitative or quantitative change in tubulin composition, leading to a reorganization of the cytoskeleton. In collaboration with Dr. Roger Sloboda, we attempted to determine whether elongation was a consequence of increased tubulin accumulation in the cells.[28] Careful measurements of tubulin content, synthesis, and degradation before and after exposure to hormone were made. No change was found for any parameter. We did find evidence, from [³H]-colchicine-binding studies, which led us to speculate that intracellular tubulin exists in two forms: one capable of binding colchicine and assembling into microtubules; and a second unable to bind colchicine, or assemble. This nonfunctional pool, we believe, is associated with recently discovered "endogenous colchicine", a factor that binds tubulin subunits and prevents microtubule assembly. Thus, we propose that one effect of ecdysterone is to alter the equilibrium between the two pools.

More recently, Sobrier et al.[29] have shown that ecdysterone actually induces the *de novo* synthesis of a β-subunit of tubulin normally absent in growing cells. In *Drosophila* there are four genes each for α- and β-tubulin, and their expression is stage and tissue specific. The β-3 isoform induced in cells appears to be a form found only in 8- to 13-h embryos *in vivo*. This period of embryonic development is a peak in terms of ecdysterone titer.

3. Aggregation Proteins

The aggregation of motile cells has been correlated with the appearance of over 30 new cell-surface glycoproteins, some of which are thought to be involved in the aggregation process. The most compelling evidence for this association comes from the work of Rickoll. He has produced both monoclonal and polyclonal antibodies specific for several hormone-dependent cell surface glycoproteins from S3 cells. Many of the same proteins also appear on the surface of imaginal disk cells *in vivo* and are ecdysterone inducible in culture. One hormone-dependent 110-Kd glycoprotein is of special interest. It is the major hormone-dependent *N*-acetylglucosamine incorporating glycoprotein seen in S3 cells, while in non-aggregating line L3 cells it is absent. This protein is secreted and displays cell binding properties. In addition, Monensin, an inhibitor of Golgi-dependent secretion, leads to reversible disaggregation, again implying a function for the 110-kDa protein in aggregation.

C. Enzymes

Ecdysterone also leads to the induction of several enzyme activities in responsive cells. These include acetylcholinesterase (AChE),[30-32] catalase,[33] dopa-decarboxylase,[34] β-galactosidase,[35] and superoxide dismutase. Typically, enzyme induction is delayed and becomes apparent only after 12 to 24 h, so that it is not clear whether any of these are actually a primary response to hormone. The availability of cloned genes will resolve this question. The pattern of AChE induction has been studied extensively and shows an interesting pattern of cell line specificity.

In Kc or MDR cells the kinetics of induction include a concentration-dependent lag period, followed by a phase in which AChE-specific activity rises linearly for at least 72 h. The hormone-insensitive MDER line derived from MDR was not inducible. S3 cells, in contrast, have been shown by us to contain a high constitutive level of AChE activity. Upon the addition of ecdysterone, AChE activity drops precipitously following a hormone concentration-dependent lag period. After this decline phase, AChE specific activity increases again at a linear rate for at least 3 d. The slope of the reinduction phase increases with increasing hormone concentration. We have found that the basal- and induced-AChE activities represent different isozymes, although the molecular basis for these different mobility forms is unknown. Parenthetically, neither of the cellular isozymes comigrate with the AChE form extracted from adult *Drosophila*. All of the AChE activities we have studied show sensitivity to eserine. One intriguing aspect of all this is that only a single AChE structural gene has been identified.

Lysate mixing experiments have been carried out. The results demonstrate that the initial disappearance of AChE activity during the decline phase is not the result of an AChE inhibitor appearing transiently. Experiments employing inhibitors of protein synthesis have indicated that the basal-AChE-specific activity in S3 cells is normally short lived (half-life about 8 h). This was true whether AChE-specific activity was determined as a function of total cell protein or as a function of isocitrate dehydrogenase activity in the same lysate. Thus, the initial decline in specific activity could be explained: (1) as a translational effect initiated by hormone (AChE mRNA is no longer used), (2) as transcriptional repression, if AChE mRNA, too, were normally short lived, or (3) it could reflect degradation or secretion of acetylcholinesterase from the cells. The induction phase in S3 cells requires the continued presence of hormone. When ecdysterone is withdrawn, the level of acetylcholinesterone declines.

D. Other Proteins

Another approach to determining changes in gene expression in response to hormone treatment has been to label proteins with [^{35}S]methionine before and after hormone treatment, to separate extracted proteins by 2-D PAGE, and then to identify changes in the patterns of labeled proteins. A number of qualitative and quantitative changes have been noticed after hormone treatment. Savakis et al.[43-45] discovered a set of ecdysterone-inducible polypeptides (EIPs) with molecular weights of about 28, 29, and 40 kDa in Kc cells. EIPs are synthesized at a low level in untreated cells, but synthesis is induced to detectable levels within 1 h after hormone treatment and to tenfold after 4 to 8 h. The increased protein levels are the result of increased mRNA production. The increase in EIP mRNA levels occurs in the presence of inhibitors such as cycloheximide, supporting the idea that EIP induction is a primary hormone response. *In situ* hybridization studies of *Drosophila* salivary gland polytene chromosomes with cDNA clones of the coding regions of the EIP genes have localized the EIP 40 gene to region 55B-D, and the EIP 28 and EIP 29 genes to a single band in the region 71C-D. Using the same approach, we discovered a set of proteins whose relative rate of synthesis increased two- to fivefold following ecdysterone addition. Induction is greatest in the S3 line cells and is evident in labeled amino acid incorporation studies by 2 to 3 h. Induction follows a dose response curve characteristic of the physiological situation, with half-maximal induction at 10^{-8} M.[46] Evidence discussed later established that these four proteins are a subset of the so-called heat shock proteins and that induction by heat shock and ecdysterone occurs by independent mechanisms. Before presenting that evidence, some discussion of the heat shock response in cultured cells is in order.

III. THE HEAT SHOCK RESPONSE

When *Drosophila* cells are exposed to temperatures above 33°C or to noxious chemicals,[47] they initiate the heat shock response. This includes a rapid and abrupt switch in the pattern of gene transcription,[48] a block in pre-mRNA splicing, and a switch from the translation of preexisting mRNA to the preferential utilization of mRNA made during heat shock.[49-52] There are numerous reviews on the topic of heat shock,[48,53-55] so that we will simply summarize the events that are pertinent to the discussion that follows.

Within a minute of heat shock, the transcription of genes active at 25°C ceases. A set of seven genes, the heat shock genes, abruptly becomes active. These genes encode the heat shock proteins: hsp 70 (5 genes at 2 loci), hsp 68, hsp 83, and the four small hsps — hsp 22, hsp 23, hsp 26, and hsp 27 (or 28) — that are linked at cytological position 67B. Simultaneously, the cell's translational machinery undergoes a switch, from the translation of 25°C mRNA to the selective and preferential translation of hsp mRNAs. The mechanism underlying these changes in transcription and translation is poorly understood. Stress somehow leads to the activation of a transcription factor required for hsp gene transcription, but the mechanism underlying the repression of all other transcription is unknown. Similarly, there is evidence that preferential translation involves some special sequence or structure found only on hsp mRNAs, but the nature of the signal is unknown.

When cells are returned to 25°C, the process reverses. However, recovery does require the production of functional heat shock proteins. That is, if heated cells are deprived of protein synthesis by inhibitors or amino acid analogs, cells returned to 25°C continue to exclusively transcribe the hsp genes. Again, the mechanism of recovery is unknown. A second interesting feature is that during recovery, the hsp mRNAs, while translatable, become exceedingly unstable and break down. Heat shock proteins apparently protect cells from the toxic effect of chemical and physical stress, but, again, the mechanism of protection is unknown.

A. Small hsps, Heat Shock Proteins, and Their Genes

Exposure of *Drosophila* cells to ecdysterone leads to the rapid synthesis and accumulation of four low molecular weight proteins. These were identified as the small hsps by three criteria.[46] First, the hormone- and heat shock-induced proteins comigrated in two-dimensional polyacrylamide gels. Second, one-dimensional peptide maps of the individual proteins from heat-shocked and hormone-treated cells were indistinguishable. Finally, the DNA clones for each of the four small hsp genes, when used as a radioactive hybridization probe, hybridized extensively to RNA found in hormone-treated cells, but not to RNA from untreated cells. Subsequent studies[19,56,57] documented the kinetics of mRNA accumulation during the two patterns of induction and showed that three other genes that are embedded in the small hsp gene cluster remain transcriptionally inactive following either treatment.

It was also important to demonstrate that the responses of ecdysterone-treated cells are specific to the hormone and not due to stress-related conditions. Three lines of evidence support the specificity of the hormone response. First, small hsp synthesis is absent in sham-treated controls, either in low ecdysterone concentrations or in high levels of mammalian steroid hormones. Second, synthesis of the small hsps in response to ecdysterone is uncoupled from synthesis of the major heat shock proteins hsp 68 and hsp 70. This situation is unique to the ecdysterone response and clearly demonstrates independent regulatory responses for each treatment. Finally, in cell lines selected for hormone insensitivity, the cells maintained a normal heat shock response but lost the ability to synthesize small hsps in response to ecdysterone.

Small hsp gene expression was soon shown to be developmentally regulated *in vivo*.[20,58-62] Expression is tissue specific (imaginal disks and oocytes) and temporally regulated (pupation and oogenesis). The physiological significance of small hsp synthesis during development is unclear, but there is some evidence that these proteins can confer thermotolerance[9] and may act as a buffer during development, protecting against environmental stress.

B. A Molecular Approach to Cells

At the molecular level, we have proposed that both ecdysterone and heat shock regulation occur via *cis*-acting regions upstream (5′) from the transcription start site of each gene. Simplistically, one could speculate that a regulatory switch, a *trans*-acting protein factor, would interact with the *cis*-acting sequences to achieve coordinate regulation. Elucidation of these regulatory regions can be approached with both genetic and biochemical experiments. A genetic approach would involve generating mutants which disrupt small hsp function and screening for altered sequences in the 5′-flanking regions. However, while the small hsp genes have been cloned and sequenced,[63-69] there is currently no assay for small hsp function. A second method would be to study sequence homologies found between gene families. Although this method has been successful in many other systems, to date only the small hsp gene family has been characterized as ecdysterone responsive. A promising functional approach is the use of DNA-mediated gene transfer. In this assay, cloned genes, when introduced into responsive cells[70-72] or into embryos[73-75] by transformation, acquire normal regulation. Then the intact genes are mutated *in vitro*, using DNA cutting enzymes, and reintroduced into cells. The loss of regulated expression, occurring in response to a specific modification of DNA, defines the affected region as a regulatory element. Two transfer systems have been successful: a transient expression assay and a transformation system. We have used the transient expression assay to define *cis*-acting sequences involved in hormone and heat shock regulation. Exogenous DNA introduced into *Drosophila* cells was taken up quite efficiently and found to be regulated in a normal pattern. To differentiate between the expression of exogenously introduced heat shock genes and their endogenous counterparts, we linked a reporter gene to 5′-flanking regions of each small heat shock gene to form chimeric clones. A series of deletions in the 5′ upstream flanking region of the chimeric clones was constructed, then assayed for heat shock or hormone inducibility. Two important results were obtained. First, clones with deletions up to a critical point retained hormone or heat shock inducibility, while deletions further downstream (3') from this point, resulted in completely inactive constructs. One can conclude that a regulatory region exists distal to the critical deletion point of each gene. Second, the critical deletion point for heat shock inducibility differed from that of ecdysone stimulation in all the small heat shock genes, supporting evidence for independent regulatory mechanisms for the two treatments. For a summary of the regulatory sequences defined for the small heat shock genes by deletional analysis,[76-94] see Figure 1.

C. Prospects

Based on the functional analysis of intact and mutated heat shock protein genes, regulatory regions involved with heat shock- and ecdysterone-induced transcription have been identified. The heat shock element is a tandem pair of 14 base pair palindromes, known as Pelham boxes. This element, or a region nearby, also appears to become preferentially sensitive to DNase digestion, upon heat shock induction.[95-99] Recently, several groups have identified a protein, known as the heat shock factor (HSF) which binds to the Pelham boxes in a cooperative and sequence-specific manner. The HSF presumably functions *in vivo* as a *trans*-acting, positive control element.[100-107]

While an ecdysterone receptor protein has been identified and partially character-ized,[108-114] it has not yet been purified. The prediction is that the ecdysterone-receptor complex like HSF functions as a sequence-specific, DNA-binding protein, the function of which is to bind to the ecdysterone-regulatory element and somehow modulate gene transcription. Since ecdysterone-receptor complexes appear to also bind to genes whose transcription is repressed, it is not clear whether the hormone-receptor complex can act as both a positive and a negative, diffusible transcription factor.

In conclusion, the *Drosophila* cell culture system has become an important tool for the molecular genetic analysis of ecdysterone action. The discovery that the small hsp gene cluster is under dual regulation adds a new dimension to this analysis. The technologies of DNA-mediated gene transfer now allow us to ask precise questions about the nature of gene regulation and to test the function of genes modified *in vitro*.

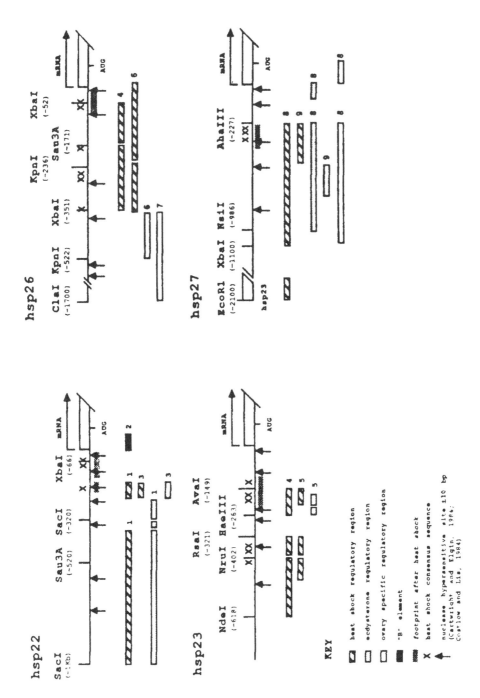

FIGURE 1. The ecdysterone and heat shock regulatory elements near the four small hsp genes, defined by functional analyses in flies and cultured cells.

REFERENCES

1. **Ashburner, M.,** Sequential gene activation by ecdysone in polytene chromosomes of *Drosophila, Dev. Biol.,* 39, 141, 1974.
2. **Echalier, G. and Ohanessian, A.,** *In vitro* culture of *Drosophila melanogaster* embryonic cells, *In Vitro,* 6, 162, 1970.
3. **Schneider, I.,** Cell lines derived from late embryonic stages of *Drosophila melanogaster, J. Embryol. Exp. Morphol.,* 27, 353, 1972.
4. **Rosset, R.,** Effects of ecdysterone on a *Drosophila* cell line, *Exp. Cell Res.,* 11, 31, 1978.
5. **Berger, E., Ringler, R., Alahiotis, S., and Frank, M.,** Ecdysone induced changes in morphology and protein synthesis in *Drosophila* cells, *Dev. Biol.,* 62, 498, 1978.
6. **Courgeon, A. M.,** Action of insect hormones at the cellular level, *Exp. Cell Res.,* 74, 327, 1972.
7. **Courgeon, A. M.,** Effect of α- and β-ecdysone on *in vitro* diploid cell multiplication in *Drosophila melanogaster, Nature (London) New Biol.,* 238, 250, 1972.
8. **Berger, E. M., Frank, M., and Abell, M. C.,** Ecdysone induced changes in protein synthesis in embryonic *Drosophila* cells in culture, in *Invertebrate Systems in Vitro,* Kurstak, E., Maramorosch, D., Eds., Elsevier/North-Holland, Amsterdam, 1980, 195.
9. **Berger, E. M. and Woodward, M. P.,** Small heat-shock proteins in *Drosophila* confer thermal tolerance, *Exp. Cell Res.,* 147, 437, 1983.
10. **Cherbas, L., Yong, C. D., Cherbas, P., and Williams, C. M.,** The morphological response of Kc-H cells to ecdysteroids: hormonal specificity, *Wilhelm Roux Arch. Entwicklungsmech. Org.,* 189, 1, 1980.
11. **Wyss, C.,** Juvenile hormone analogue counteracts growth stimulation and inhibition of ecdysones in clonal *Drosophila* cell line, *Experientia,* 32, 1272, 1976.
12. **Wyss, C. and Eppenberger, H.,** Morphological and proliferative response of Schneider's *Drosophila* cell line 3 to ecdysterone, *Experientia,* 34, 961, 1978.
13. **Vitek, M. and Berger, E.,** Transcriptional and post-transcriptional regulation of small heat shock protein synthesis, *J. Mol. Biol.,* 178, 173, 1984.
14. **Stevens, B. and O'Connor, J. D.,** The acquisition of resistance to ecdysteroids in cultured *Drosophila* cells, *Dev. Biol.,* 94, 176, 1982.
15. **Stevens, B., Alvarez, C. M., Bohman, R., and O'Connor, J. D.,** An ecdysteroid-induced alteration in the cell cycle of cultured *Drosophila* cells, *Cell,* 22, 675, 198.
16. **Wyss, C.,** Cell hybrid analysis of ecdysone sensitivity and resistance in *Drosophila* cell lines, in *Invertebrate Systems in Vitro,* Kurstak, E., Maramorosch, K., and Dübendorfer, A., Eds., Elsevier/North-Holland, Amsterdam, 1980, 279.
17. **Wyss, C.,** Transformation of a mutant *Drosophila* cell line *in vitro, Experientia,* 37, 665, 1981.
18. **Maroy, P., Dennis, R., Beckers, C., Sage, B. A., and O'Connor, J. D.,** Demonstration of an ecdysteroid receptor in a cultured cell line of *Drosophila melanogaster, Proc. Natl. Acad. Sci. U.S.A.,* 75, 6035, 1978.
19. **Ireland, R., Berger, E., Sirotkin, K., Yund, M., Osterbur, D., and Fristrom, J.,** Ecdysterone induces the transcription of four heat-shock genes in *Drosophila* S3 cells and imaginal discs, *Dev. Biol.,* 93, 498, 1982.
20. **Couderc, J. L. and Dastugue, B.,** Ecdysterone induced modifications of protein synthesis in a *Drosophila melanogaster* cultured cell line, *Biochem. Biophys. Res. Commun.,* 97, 173, 1980.
21. **Couderc, J. L., Codic, A. L., Sobrier, M. L., and Dastugue, B.,** Ecdysterone induction of actin synthesis and polymerization in a *Drosophila* cultured cell line, *Biochem. Biophys. Res. Commun.,* 107, 188, 1982.
22. **Couderc, J. L., Sobrier, M. L., Giraud, G., Becker, J. L., and Dastugue, B.,** Actin gene expression is modulated by ecdysterone in *Drosophila melanogaster* cells, *J. Mol. Biol.,* 164, 419, 1983.
23. **Berger, E. M., Cox, G., Ireland, R., and Weber, L.,** Actin content and synthesis in differentiating *Drosophila* cells in culture, *J. Insect Physiol.,* 27, 129, 1981.
24. **Vitek, M. P., Morganelli, C. M., and Berger, E. M.,** Stimulation of cytoplasmic actin gene transcription and translation in cultured *Drosophila* cells by ecdysterone, *J. Biol. Chem.,* 259, 1738, 1984.
25. **Berger, E. and Cox, G.,** A precursor of cytoplasmic actin in cultured *Drosophila* cells, *J. Cell Biol.,* 81, 680, 1979.
26. **Berger, E., Cox, G., Weber, L., and Kenny, S.,** Actin acetylation in *Drosophila* tissue culture cells, *Biochem. Genet.,* 19, 321, 1981.
27. **Fyrberg, E. A., Bond, B. J., Hershey, N., Mixter, K. S., and Davidson, N.,** The actin genes of *Drosophila* protein coding regions are highly conserved but intron positions are not, *Cell,* 24, 107, 1981.
28. **Berger, E., Sloboda, R. D., and Ireland, R.,** Tubulin content and synthesis in differentiating cells in culture, *Cell Motility,* 1, 113, 1980.
29. **Sobrier, M., Couderc, J. L., Chapel, S., and Dastugue, B.,** Expression of a new β tubulin subunit is induced by 20-hydroxyecdysone in *Drosophila* cultured cells, *Biochem. Biophys. Res. Commun.,* 134, 191, 1986.

30. **Cherbas, P., Cherbas, L., and Williams, C. M.,** Induction of acetylcholinesterase activity by β-ecdysone in a *Drosophila* cell line, *Science,* 197, 275, 1977.

31. **Berger, E. and Wyss, C.,** Acetylcholinesterase induction by β-ecdysone in *Drosophila* cells and their hybrids, *Somatic Cell Genet.,* 6, 631, 1980.

32. **Best-Belpomme, M. and Courgeon, A. M.,** Ecdysterone and acetylcholinesterase activity in cultured *Drosophila* cells, *FEBS Lett.,* 82, 345, 1987.

33. **Best-Belpomme, M. and Ropp, M.,** Catalase is induced by ecdysterone and ethanol in *Drosophila melanogaster* cells, *Eur. J. Biochem.,* 121, 349, 1982.

34. **Spencer, C. A., Stevens, B., O'Connor, J. D., and Hodgetts, R. B.,** A novel form of dopa decarboxylase produced in *Drosophila* cells in response to 20-hydroxyecdysone, *Can. J. Biochem. Cell Biol.,* 61, 818, 1983.

35. **Best-Belpomme, M., Courgeon, A. M., and Rambach, A.,** Beta-galactosidase is induced by hormone in *Drosophila melanogaster* cell cultures, *Proc. Natl. Acad. Sci. U.S.A.,* 75, 6102, 1978.

36. **Johnson, T. K., Brown, A. and Dennell, R.,** Changes in cell surface proteins of cultured *Drosophila melanogaster* cells exposed to 20-hydroxyecdysone, *Wilhelm Roux Arch. Dev. Biol.,* 192, 103, 1983.

37. **Dennis, R. D. and Haustein, D.,** Ecdysteroid related changes in cell surface properties of a *Drosophila* cell line, *Insect Biochem.,* 12, 83, 1982.

38. **Yudin, A. I., Clark, W. H., and Chang, E. S.,** Ecdysteroid induction of cell surface contacts in a *Drosophila* cell line, *J. Exp. Zool.,* 219, 399, 1982.

39. **Woods, D. F. and Poodry, C. A.,** Cell surface proteins of *Drosophila.* I. Changes induced by 20-OH-ecdysone, *Dev. Biol.* 96, 23, 1983.

40. **Woods, D., Rickoll, W. L., Birr, C., Poodry, C. A., and Fristrom, J. W.,** Alterations in the cell surface proteins of *Drosophila* during morphogenesis, *Wilhelm Roux Arch. Dev. Biol.,* in press.

41. **Rickoll, W. L., Stachowiak, D. A., Galewsky, S., Junio, M. A., and Hayes, E. S.,** Differential effects of 20-hydroxyecdysone on cell interactions and surface proteins in *Drosophila* cell lines, *Insect Biochem.,* 16, 211, 1986.

42. **Rickoll, W. L. and Galewsky, S.,** Antibodies recognizing 20-hydroxyecdysone-dependent cell surface antigens during morphogenesis in *Drosophila, Wilhelm Roux Arch. Dev. Biol.,* in press.

43. **Savakis, C., Demetri, G., and Cherbas, P.,** Ecdysteroid inducible polypeptides in a *Drosophila* cell line, *Cell,* 22, 665, 1980.

44. **Savakis, C., Koehler, M., and Cherbas, P.,** cDNA clones for the ecdysone-inducible polypeptide (EIP) mRNAs in *Drosophila* Kc cells, *EMBO J.,* 3, 235, 1984.

45. **Cherbas, L., Schulz, R., Koehler, M., Savakis, C., and Cherbas, P.,** Structure of the Eip 28/29 gene, an ecdysone-inducible gene from *Drosophila, J. Mol. Biol.,* 189, 617, 1986.

46. **Ireland, R. and Berger, E.,** Synthesis of low molecular weight heat shock peptides by ecdysterone in a cultured *Drosophila* cell line, *Proc. Natl. Acad. Sci. U.S.A.,* 79, 855, 1982.

47. **Courgeon, A., Maisonhaute, C., and Best-Belpomme, M.,** Heat shock proteins are induced by cadmium in *Drosophila* cells, *Exp. Cell Res.,* 153, 515, 1984.

48. **Ashburner, M. and Bonner, J. J.,** The induction of gene activity in *Drosophila* by heat shock, *Cell,* 17, 241, 1979.

49. **DiDomenico, B., Bugalsky, G. E., and Lindquist, S.,** The heat shock response is self-regulated at both the transcriptional and postcriptional levels, *Cell,* 31, 593, 1982.

50. **DiDomenico, B. J., Bugalsky, G. E., and Lindquist, S.,** Heat shock and recovery are mediated by different translational mechanisms, *Proc. Natl. Acad. Sci. U.S.A.,* 79, 6181, 1982.

51. **Storti, R. V., Scott, M. P., Rich, A., and Pardue, M. L.,** Translational control of protein synthesis in response to heat shock in *Drosophila melanogaster* cells, *Cell,* 22, 825, 1985.

52. **Klemenz, R., Hultmark, D., and Gehring, W. J.,** Selective translation of heat shock mRNA in *Drosophila melanogaster* depends on sequence information in the leader, *EMBO J.,* 4, 2053, 1985.

53. **Schlesinger, M., Ashburner, M., and Tissieres, A., Eds.,** *Heat Shock, From Bacteria to Man,* Cold Spring Harbor Laboratory, Cold Spring Harbor, NY, 1982.

54. **Neidhardt, F. C., Van Bogelen, R. A., and Vaughn, V.,** The genetics and regulation of heat-shock proteins, *Annu. Rev. Genet.,* 18, 295, 1984.

55. **Munro, S. and Pelham, H. R. B.,** What turns on heat shock genes?, *Nature (London),* 317, 477, 1985.

56. **Berger, E. M., Vitek, M., and Morganelli, C. M.,** Transcript length heterogeneity at the small heat shock protein genes of *Drosophila, J. Mol. Biol.,* 186, 137, 1985.

57. **Vitek, M. P. and Berger, E. M.,** Transcriptional and post-transcriptional regulation of small heat shock protein synthesis, *J. Mol. Biol.,* 178, 173, 1984.

58. **Zimmerman, J. L., Petri, W., and Meselson, M.,** Accumulation of a specific subset of *D. melanogaster* heat shock protein mRNAs in normal development without heat shock, *Cell,* 32, 1161, 1982.

59. **Cheney, C. N. and Shearn, A.,** Developmental regulation of *Drosophila* imaginal disc proteins; synthesis of heat shock protein under non-heat shock conditions, *Dev. Biol.,* 95, 325, 1983.

60. **Sirotkin, K. and Davidson, N.,** Developmentally regulated transcription from *Drosophila melanogaster* chromosomal site 67B, *Dev. Biol.,* 89, 196, 1982.

61. **Mason, P. J., Hall, L. M. C., and Gausz, J. C.,** The expression of heat shock genes during normal development in *Drosophila melanogaster, Mol. Gen. Genet.,* 194, 73, 1984.
62. **Glaser, R. L., Wolfner, M. F., and Lis, J. T.,** Spatial and temporal pattern of hsp 26 expression during normal development, *EMBO J.,* 4, 747, 1986.
63. **Craig, E. A. and McCarthy, B. J.,** Four *Drosophila* heat shock genes at 67B: characterization of recombinant plasmids, *Nucleic Acid Res.,* 8, 4441, 1980.
64. **Craig, E. A., McCarthy, B. J., and Wadsworth, S. C.,** Sequence organization of two recombinant plasmids containing genes for the major heat shock-induced protein of *D. melanogaster, Cell,* 16, 575, 1979.
65. **Ingolia, T. D. and Craig, E. A.,** Primary sequence of the 5' flanking regions of the *Drosophila* heat shock genes in chromosome subdivision 67B, *Nucleic Acid Res.,* 9, 1627, 1981.
66. **Ingolia, T. D. and Craig, E. A.,** Four small *Drosophila* heat shock proteins are related to each other and to mammalian α-crystallin, *Proc. Natl. Acad. Sci. U.S.A.,* 79, 2360, 1982.
67. **Corces, V., Holmgren, R., Freund, R., Morimoto, R., and Meselson, M.,** Four heat shock proteins of *Drosophila melanogaster* coded within a 12 kilobase region in chromosome subdivision 67B, *Proc. Natl. Acad. Sci. U.S.A.,* 77, 5390, 1980.
68. **Voellmy, R., Goldschmidt-Clermont, M., Southgate, R., Tissieres, A., Levis, R., and Gehring, W. J.,** A DNA segment isolated from chromosomal site 67B in *Drosophila melanogaster* contains four closely linked heat shock genes, *Cell,* 23, 261, 1981.
69. **Southgate, R., Ayme, A., and Voellmy, R.,** Nucleotide sequence analysis of the *Drosophila* small heat shock gene cluster at locus 67B, *J. Mol. Biol.,* 165, 35, 1983.
70. **DiNocera, P. and Dawid, I. B.,** Transient expression of genes introduced into cultured cells of *Drosophila, Proc. Natl. Acad. Sci. U.S.A.,* 80, 7095, 1983.
71. **Burke, J. F., Sinclair, J. H., Sang, J. H., and Ish-Horowicz, D.,** An assay for transient gene expression in transfected *Drosophila* cells, using [³H] guanine incorporation, *EMBO J.,* 3, 2549, 1984.
72. **Burke, J. F., Pinchin, S. M., Ish-Horowicz, D., Sinclair, J. H., and Sang, J. H.,** Integration of heat shock genes transfected into cultured *Drosophila melanogaster* cells, *Somatic Cell. Mol. Genet.,* 10, 579, 1984.
73. **Rubin, G. M. and Spradling, A. C.,** Genetic transformation of *Drosophila* with transposable element vectors, *Science,* 218, 348, 1982.
74. **Rubin, G. M. and Spradling, A. C.,** Vectors for P-element-mediated gene transfer in *Drosophila, Nucleic Acids Res.,* 11, 1983.
75. **Spradling, A. C. and Rubin, G. M.,** Transposition of cloned P-elements into *Drosophila* germ-line chromosomes, *Science,* 281, 341, 1982.
76. **Hoffman, E. P. and Corces, V. G.,** Correct temperature induction and developmental regulation of a cloned heat shock gene transformed into the *Drosophila* germ line, *Mol. Cell. Biol.,* 4, 2883, 1984.
77. **Hoffman, E. and Corces, V.,** Sequences involved in temperature and ecdysterone-induced transcription are located in separate regions of a *Drosophila melanogaster* heat shock gene, *Mol. Cell. Biol.,* 6, 663, 1986.
78. **Morganelli, C. M. and Berger, E.,** Transient expression of homologous genes in cultured *Drosophila* cells, *Science,* 224, 1004, 1984.
79. **Morganelli, C. M., Berger, E. M., and Pelham, H. R. B.,** Transcription of *Drosophila* small *hsp-tk* hybrid genes is induced by heat shock and by ecdysterone in transfected *Drosophila* cells, *Proc. Natl. Acad. Sci. U.S.A.,* 82, 5865, 1985.
80. **Riddihough, G. and Pelham, H. R. B.,** Activation of the *Drosophila* hsp 27 promoter by heat shock and by ecdysterone involves independent and remote regulatory sequences, *EMBO J.,* 5, 1653, 1986.
81. **Hultmark, D., Klemenz, R., and Gehring, W. J.,** Translational and transcriptional control elements in the untranslated leader of the heat-shock gene hsp 22, *Cell,* 44, 429, 1986.
82. **Klemenz, R. and Gehring, W. J.,** Sequence requirement for the expression of the *Drosophila melanogaster* heat shock protein hsp 22 gene during heat shock and normal development, *Mol. Cell. Biol.,* 6, 2011, 1986.
83. **Cohen, R. S. and Meselson, M.,** Separate regulatory elements for the heat-inducible and ovarian expression of the *Drosophila* hsp 26 gene, *Cell,* 43, 737, 1985.
84. **Lawson, R., Mestril, R., Schiller, P., and Voellmy, R.,** Expression of heat shock-β-galactosidase hybrid genes in cultured *Drosophila* cells, *Mol. Gen. Genet.,* 198, 116, 1984.
85. **Mestril, R., Rungger, D., Schiller, P., and Voellmy, R.,** Identification of a sequence element in the promoter of the *Drosophila melanogaster* hsp 23 gene that is required for its heat activation, *EMBO J.,* 4, 2971, 1985.
86. **Mestril, R., Schiller, P., Admin, J., Klapper, H., Ananthan, J., and Voellmy, R.,** Heat shock and ecdysterone activation of the *Drosophila melanogaster* hsp 23 gene; a sequence element implied in development regulation, *EMBO J.,* 5, 1667, 1986.

87. **Lawson, R., Mestril, R., Luo, Y., and Voellmy, R.,** Ecdysterone selectively stimulates the expression of a 23000-Da heat shock protein-β-galactosidase hybrid gene in cultured *Drosophila* cells, *Dev. Biol.,* 110, 321, 1985.

88. **Amin, J., Mestril, R., Lawson, R., Klapper, H., and Voellmy, R.,** The heat shock consensus sequence is not sufficient for hsp 70 gene expression in *Drosophila melanogaster, Mol. Cell. Biol.,* 5, 197, 1985.

89. **Pauli, D., Spierer, A., and Tissieres, A.,** Several hundred base pairs upstream of *Drosophila* hsp 23 and 26 genes are required for their heat induction in transformed flies, *EMBO J.,* 5, 755, 1986.

90. **Pelham, H. R. B.,** A regulatory upstream promoter element in the *Drosophila* hsp 70 heat-shock gene, *Cell,* 30, 517, 1982.

91. **Pelham, H. R. B. and Bienz, M.,** A synthetic heat-shock promoter element confers heat-inducibility on the herpes simplex virus thymidine kinase gene, *EMBO J.,* 1, 1473, 1982.

92. **Berger, E., Morganelli, C. M., and Torrey, D.,** Natural and synthetic *Drosophila* heat shock protein gene promoters, *Somat. Cell Mol. Genet.,* 12, 433, 1986.

93. **Dudler, R. and Travers, A. A.,** Upstream elements necessary for optimal function of the hsp 70 promoter in transformed flies, *Cell,* 38, 391, 1984.

94. **Amin, J., Mestril, R., Schiller, P., Dreano, M., and Voellmy, R.,** Organization of the *Drosophila melanogaster* heat shock regulation unit, *Mol. Cell. Biol.,* 7, 1055, 1987.

95. **Wu, C.,** The 5' ends of *Drosophila* heat shock genes in chromatin are hypersensitive to DNase I, *Nature (London),* 286, 854, 1980.

96. **Keene, M. A. and Elgin, S. C. R.,** Micrococcal nuclease as a probe of DNA sequence organization and chromatin structure, *Cell,* 27, 57, 1981.

97. **Elgin, S. C. R.,** DNAse I-hypersensitive sites of chromatin, *Cell,* 27, 413, 1981.

98. **Costlow, N. and Lis, J. T.,** High-resolution mapping of DNase I-hypersensitive sites of *Drosophila* heat shock genes in *Drosophila melanogaster* and *Saccharomyces cerevisiae, Mol. Cell. Biol.,* 4, 1853, 1984.

99. **Cartwright I. and Elgin, S. C. R.,** Nucleosomal instability and induction of new upstream protein-DNA associations accompany activation of four small heat shock protein genes in *Drosophila melanogaster, Mol. Cell. Biol.,* 6, 779, 1986.

100. **Craine, B. L. and Kornberg, T.,** Activation of the major *Drosophila* heat shock genes in vitro, *Cell,* 25, 671, 1981.

101. **Jack, R. S., Gehring, W., and Brack, C.,** Protein component from *Drosophila* larval nuclei showing sequence specificity for a short region near a major heat-shock protein gene, *Cell,* 24, 321, 1981.

102. **Wu, C.,** Two protein-binding sites in chromatin implicated in the activation of heat shock genes, *Nature (London),* 309, 229, 1984.

103. **Wu, C.,** Activating protein factor binds in vitro to upstream control sequences in heat shock gene chromatin, *Nature (London),* 311, 81, 1984.

104. **Parker, C. and Topol, J.,** A *Drosophila* RNA polymerase II transcription factor binds to the regulatory site of an hsp 70 gene, *Cell,* 37, 273, 1984.

105. **Topol, J., Ruden, D. M., and Parker, C. S.,** Sequences required for in vitro transcriptional activation of a *Drosophila* hsp 70 gene, *Cell,* 42, 527, 1985.

106. **Shuey, D. J. and Parker, C.,** Bending of promoter DNA on binding of heat shock transcription factor, *Nature (London),* 323, 459, 1986.

107. **Shuey, D. J. and Parker, C. S.,** Binding of *Drosophila* heat shock gene transcription factor to the hsp 70 promoter, *J. Biol. Chem.,* 261, 7934, 1986.

108. **Yund, M. A., King, D. S., and Fristrom, J. W.,** Ecdysteroid receptors in imaginal discs of *Drosophila melanogaster, Proc. Natl. Acad. Sci. U.S.A.,* 75, 6039, 1978.

109. **Maroy, P., Dennis, R., Becker, C., Sage, B. A., and O'Connor, J. D.,** Demonstration of an ecdysteroid receptor in a cultured cell line of *Drosophila melanogaster, Proc. Natl. Acad. Sci. U.S.A.,* 75, 6035, 1978.

110. **Sage, B. A., Tanis, M. A., and O'Connor, J. D.,** Characterization of ecdysteroid receptors in cytosol and naive nuclear preparations of *Drosophila* Kc cells, *J. Biol. Chem.,* 257, 6373, 1982.

111. **Gronemeyer, H., Hameister, H., and Pongs, O.,** Photoinduced bonding of endogenous ecdysterone to salivary gland chromosomes of *Chironomus tentans, Chromosoma,* 82, 543, 1981.

112. **Gronemeyer, H. and Pongs, O.,** Localization of ecdysterone on polytene chromosomes of *Drosophila melanogaster, Proc. Natl. Acad. Sci. U.S.A.,* 77, 2108, 1980.

113. **Schaltmann, K. and Pongs, O.,** A simple procedure for blotting of proteins to study antibody specificity and antigen structure, *Hoppe Seylers Z. Physiol. Chem.,* 361, 207, 1980.

114. **Schaltmann, K. and Pongs, O.,** Identification and characterization of the ecdysterone receptor in *Drosophila melanogaster* by photoaffinity labeling, *Proc. Natl. Acad. Sci. U.S.A.,* 79, 6, 1982.

Chapter 19

GENE TRANSFER VECTORS OF A BACULOVIRUS, *BOMBYX MORI* NUCLEAR POLYHEDROSIS VIRUS, AND THEIR USE FOR EXPRESSION OF FOREIGN GENES IN INSECT CELLS

Susumu Maeda

TABLE OF CONTENTS

I. INTRODUCTION

Recent advances in biotechnology have allowed the introduction of foreign genes into heterologous hosts. Plasmid or phage vectors are widely used vehicles which can transfer foreign genes into *Escherichia coli* or other microorganisms. These vectors can also be used for expression of foreign genes for production of specific gene products.

A variety of host vector systems have been explored to express genes of eukaryotic organisms. Although only a few plasmid-like DNAs are found in the animal kingdom, many viral vectors have been widely tested. Nuclear polyhedrosis virus (NPV) belongs to the family Baculoviridae, is widely distributed in nature,[1] and has many advantageous characteristics as a vector. Expression vector systems using *Autographa californica* NPV (AcNPV)[2,3] and *Bombyx mori* NPV (BmNPV)[4,5] have been established. These Baculovirus expression systems are now considered as the most interesting vectors for high-level expression of foreign genes. BmNPV, which has been used for the construction of our vectors, has an additional advantage in that it expands the foreign gene expression to an *in vivo* system, the larvae of the silkworm.

In this chapter, I would like to describe many new transfer vectors of BmNPV and the use of such vectors in expression experiments. The expression of foreign genes *in vitro*, i.e., in established cell lines, is the predominant subject of this discussion. Expression of foreign genes *in vivo*, in silkworm larvae, will be described in detail elsewhere.[6]

II. CHARACTERISTICS OF *B. MORI* NUCLEAR POLYHEDROSIS VIRUS

A. Life Cycle of a Nuclear Polyhedrosis Virus

Insect pathogenic viruses are divided into seven families. Viruses of the most commonly found group, the family Baculoviridae, are characterized by a double-stranded circular DNA genome within a rod-shaped, enveloped virion.[1]

The subfamily, NPV, is found in several orders of insects, mainly in lepidopterans,[1] and has a unique characteristic producing proteinaceous inclusion bodies at a late stage of infection. These polyhedral inclusion bodies contain mature virions which cause horizontal transmission of the virus to other larvae by dispersion from an infected, dead body. Virions released from occlusion bodies during digestion in the midgut of the host cause the primary infection. The secondary infection is caused by a budded virion produced by these initially infected cells. At a late stage of infection, a large number of inclusion bodies containing mature virions are produced (Figure 1). These inclusion bodies are stable and can protect progeny virions from inactivating agents found in the environment.

B. Characteristics of the Polyhedrin Gene of *B. mori* Nuclear Polyhedrosis Virus

The polyhedral protein accumulates up to 20 to 30% of the total protein of infected cells at a late stage of infection. *In vitro* translation experiments of extracted messenger RNA (mRNA) reveal that more than 90% of total mRNA encodes the polyhedrin protein.[7] The polyhedrin gene of BmNPV is located in a 3.9-kb *Hind*III fragment in the 130-kb viral genome.[5] The DNA sequence of this fragment and the deduced amino acid sequence are well conserved compared to polyhedrin genes of other NPVs.[8]

It is known that the promoter of the polyhedrin gene is controlled by early gene products because the transfection of a cloned plasmid DNA containing the polyhedrin gene with 5' and 3' flanking regions does not give any expression of the gene. This may depend on a highly conserved 15-base-long sequence found near the transcriptional start site of the polyhedrin gene and another late gene, p10, in several NPVs.[8]

C. Advantageous Characters of *B. mori* Nuclear Polyhedrosis Virus as a Vector

NPV has several advantageous characters which make it a good vector. The virus has (1)

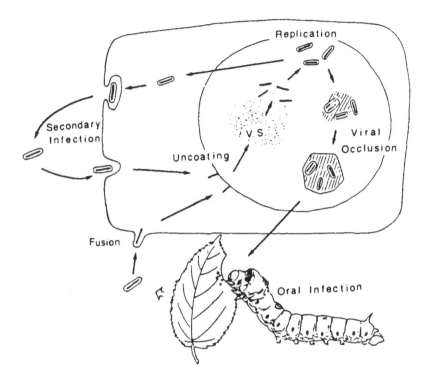

FIGURE 1. Life cycle of the *Bombyx mori* nuclear polyhedrosis virus (BmNPV).

a circular double-stranded genome, which is easily treated with restriction enzymes etc.; (2) a rod-shaped capsid which might allow it to contain an extra DNA fragment; (3) a cell line which is susceptible to viral infection; and (4) the polyhedrin gene suitable for insertion of foreign genes for the following reasons: the polyhedrin gene is nonessential for viral reproduction, is a late gene which has a strong promoter, and is a marker for insertion of a foreign gene, detected easily by light microscopy.

To establish the vector systems, it is essential to obtain a susceptible cell line which can be used for plaque assays and also to isolate the polyhedrin gene into a plasmid or phage.

III. ESTABLISHED CELL LINES FROM THE SILKWORM, *B. MORI*

A. Cell Lines Established from *B. mori*

The first insect cell line was established by Grace[9] in 1962, and several hundred additional cell lines have been established to date. Cell lines which are susceptible to insect virus infection are extremely important for the study of viral replication. Our knowledge of mammalian viruses rapidly accumulated after the establishment of cell lines for cloning and titration. AcNPV, which has a wide host range *in vivo*, i.e., in insects, also has a wide host spectrum in established cell lines. After a plaque assay system for AcNPV was developed using SeaPlaque agarose,[10] many molecular studies were performed, and AcNPV has become a model system for replication of all baculoviruses.

In contrast to AcNPV, very few cell lines are susceptible to infection with other nuclear polyhedrosis viruses. In the case of BmNPV, more than 30 insect cell lines tested are nonpermissive. The BmN cell line[11] obtained from Dr. L. Volkman was the only cell line susceptible to BmNPV when tested in 1981. BmN cells produce a large number of polyhedra and show a typical cytopathic effect after infection.[12] Although there is another report of establishment of a cell line from the silkworm, *B. mori*, which supports the replication of BmNPV,[13] that cell line is not available.

Table 1
FORMULA OF TC-100 1000 ×
VITAMIN MIXTURE

Component	Amount (mg/100 ml)
Thiamine HCl	2.0
Riboflavin	2.0
Calcium pantothenate	2.0
Pyridoxine HCl	2.0
p-Aminobenzoic acid	2.0
Folic acid	2.0
Nicotinic acid	2.0
Inositol	2.0
Biotin	1.0
Choline HCl	20.0

More recently, there have been several available cell lines which are susceptible to BmNPV. BoMo, established from embryos of a strain of the silkworm Kuroko, produces a large number of polyhedra in infected nuclei.[14] At least two other susceptible cell lines from different strains of *B. mori* have been established.[15] The Bm36 cell line[16] also supports the replication of BmNPV, but polyhedral inclusion bodies are not clearly identified even at a late stage of infection.[17] Furthermore, DNA analysis of Bm36 suggests that the origin of the cell might be an insect other than *B. mori*. BmN is confirmed as *B. mori* origin by the same method.[18]

For the studies of viral replication, a subline from BmN which shows clear plaque formation during infection of occlusion-negative BmNPV has been cloned. One of the 20 clones (BmN-4) from BmN is used for expression experiments because of the ease of isolation of a recombinant virus by showing a clear cytopathic effect.[17]

B. Cell Culture of BmN Cell Line

BmN cells (BmN-4) can be maintained by methods similar to those used for other insect cell lines. The appropriate temperature for subculture is around 27°C. The culture medium is TC-100 (a modified TC-10),[19] supplemented with 10% fetal calf serum (FCS). The composition is shown in Tables 1 and 2. Although the medium is commercially available, it is difficult to keep the cells in good condition using the commercial medium. Since there is some variation in the quality of different lots of FCS, it is recommended to select a good batch of FCS by testing cell morphology and growth.

Doubling time of BmN cells is comparatively lower than other cell lines. Furthermore, as BmN cells stick strongly to the surface of flasks, fairly large quantities of cells are lost during passages. Trypsin treatment is harmful to BmN as in other insect cell lines. Strong pipetting also causes damage to the cells. The recommended procedure for subculture is as follows. After discarding the old medium, confluent cells are suspended in fresh complete medium with a rubber policeman, pipetted gently to dissociate aggregates of cells, and seeded into fresh flasks at a dilution ratio of 1:2. The cells are generally passaged every 4 to 5 d. When a lot of dead cells are found in the medium, replacement with fresh medium is recommended.

BmN cells can be stored for at least 6 months at −70°C or in liquid nitrogen without loss of viability. Cells harvested from confluent cultures are suspended in fresh medium containing 10% FCS on ice at a cell density of 2×10^6 cells per milliliter. The cell suspension is mixed with a 10% volume of dimethyl sulfoxide (DMSO), and 1-ml aliquots are pipetted into 2-ml screwcapped plastic tubes. These tubes are wrapped with two to three layers of paper towels to control the rate of temperature decrease and are transferred to a deep freezer.

Table 2
FORMULA OF TC-100

	Component	Amount/4 l	Amount/1 l
(A)	KCl	11.48 g	2.87 g
	CaCl$_2$·2H$_2$O	5.28 g	1.32 g
	MgCl$_2$·6H$_2$O	9.12 g	2.28 g
	MgSO$_4$·7H$_2$O	11.12 g	2.78 g
	Glucose	4.0 g	1.0 g
	Tryptose broth	10.04 g	2.51 g
	L-Alanine	0.9 g	0.23 g
	β-Alanine	0.8 g	0.2 g
	L-Arginine HCl	2.8 g	0.7 g
	L-Aspartic acid	1.4 g	0.35 g
	L-Asparagine	1.6 g	0.4 g
	L-Glutamic acid	2.4 g	0.6 g
	L-Glutamine	2.4 g	0.6 g
	Glycine	2.6 g	0.65 g
	L-Histidine	10.0 g	2.5 g
	L-Isoleucine	0.2 g	50 mg
	L-Leucine	0.3 g	75 mg
	L-Lysine HCl	2.5 g	0.625 g
	L-Methionine	0.2 g	50 mg
	L-Phenylalanine	0.6 g	0.15 g
	L-Proline	1.4 g	0.35 g
	DL-Serine	4.4 g	1.1 g
	L-Threonine	0.7 g	0.175 g
	L-Valine	0.4 g	0.1 g
(B)	L-Cystine	0.1 g	25 mg
	L-Tryptophan	0.4 g	0.1 g
	L-Tyrosine	0.2 g	50 mg
(C)	NaH$_2$PO$_4$·2H$_2$O	4.55 g	1.14 g
(D)	NaHCO$_3$	1.4 g	0.35 g
(E)	1000× vitamins	4 ml	1 ml

For stock in liquid nitrogen, the tubes can be transferred after incubation for several hours at −70°C. To recover cells from the stock, the cells in the tube are quickly thawed in a cold water bath and then seeded into a 25-cm² flask with an additional 3 to 4 ml of fresh medium. After the cells are attached to the surface, the medium is replaced with fresh medium. If unattached dead cells are found in the medium, these cells can be removed by changing the medium without passage.

C. Preparation of the Medium

TC-100 culture medium is made by the following procedure. As the medium is sterilized by filtration, viral contamination should be avoided during the preparation of medium. It is recommended that all glassware and water be sterilized by autoclave.

It is convenient to prepare a separate stock solution for concentrated (1000×) vitamins as follows. Each of the ten chemicals shown in Table 1 is dissolved in 100 ml of double-distilled water. These stock solutions can be stored at −20°C. As precipitation is usually seen in the solution even after gentle mixing, mix well before use.

To make 4 l of the complete medium, each of the chemicals listed in Table 2 (A) is added into 3 l of double-distilled water with stirring. Insoluble amino acids in Table 2 (B) are separately dissolved in 150 ml of 0.2 *N* HCl at around 50°C. NaH$_2$PO$_4$·H$_2$O and NaHCO$_3$ (Table 2 [C] and [D]) are dissolved in 100 ml each of distilled water. These separately prepared solutions and 4 ml of 1000× vitamin stock are added into the above solution.

After complete dissolution of the chemicals, pH is adjusted to 6.2 with 5 *N* KOH (usually less than 2 ml). Finally, the volume is adjusted to 4 l with distilled water. For plaque assay, 1.33 × TC-100 is prepared simply by making 750 ml using the amount of chemicals for 1 l (Table 2) of the complete medium. Higher concentration of TC-100 may cause precipitation during storage at 5°C, which will damage the cells.

The medium is sterilized by filtration (0.22-μm filter) and can be stored at 5°C for up to 4 months. Prefiltration might be helpful to prevent clogging of filters with undissolved chemicals.

D. Plaque Assay

Plaque assay is essential for the isolation of genetically pure virus as a clone and the estimation of accurate titers of biologically active viral particles. This method can be used in complementation experiments for genetic analysis of a virus and also in isolation of a mutant such as a recombinant virus.

As insect cells are generally sensitive to contaminating materials found in agar or agarose, methyl cellulose was first used for the overlay in plaque assay.[20] Purified agarose, SeaPlaque agarose (FMC Corporation) is found to be suited as an overlay material because of a low toxicity to insect cells. The overlay solution contains TC-100, 5% FCS, and 0.75% SeaPlaque agarose.

The procedure of a plaque assay of BmNPV on BmN cells is as follows.[12] A monolayer of 2 × 10^6 cells in a 60-mm dish (4 to 5 ml for culture) is used for plaque assay. The cells are seeded at least 2 h prior to plaque assay. The dish should be rocked after seeding to produce a uniform cell density. Confluent cells in a 60-mm dish can provide two 60-mm dishes for plaque assay.

For viral infection, the culture fluid is discarded (if many, aspiration is recommended), and 100 μl (or 200 μl) of the viral (usually diluted) solution is added. After incubation for 1 h with rocking at 15-min intervals, the monolayers are overlaid with TC-100 medium containing 0.75% SeaPlaque agarose and 5% FCS.

The overlay solution is prepared as follows. SeaPlaque agarose is suspended at 3% in double-distilled water in a bottle (or a glass test tube) and autoclaved. The bottle is kept at 40 to 50°C after autoclaving. The amount of the agarose and distilled water is calculated by the following equations:

$$\text{Agarose (mg)} = (\text{number of plates} + 1) \times 34 \text{ mg}$$

$$\text{Water (ml)} = (\text{number of plates} + 1) \times 1.13 \text{ ml}$$

Pre-warmed (37°C) 1.33 × TC-100 medium containing 5% FCS and 60 μg/ml of kanamycin is added into the agarose solution and mixed immediately. The amount is calculated as follows:

$$1.33 \times \text{TC-100 (ml)} = (\text{number of plates} + 1) \times 3.44 \text{ ml}$$

$$\text{FCS (ml)} = (\text{number of plates} + 1) \times 0.23 \text{ ml}$$

$$1000 \times \text{kanamycin (60 mg/ml)} = (\text{number of plates} + 1) \times 9$$

To prevent detachment of the cells, the agarose solution (4.5 ml) is pipetted slowly onto an infected monolayer. The overlaid plates are kept on a bench for 20 min with the lid slightly opened to solidify the solution. Plaques can be identified after incubation for 4 to 6 d at 27°C. The titer of the original solution is calculated by dilution rate and the volume added to the plate. To help visualize the plaques more clearly, a second agarose overlay (2 ml) containing 0.02% of neutral red is recommended. The counting of plaques should be done at least 12 h after the second overlay.

The viral solution added to the plate for titration should be diluted to give between 10 to 200 plaques per dish. The viral titer in infected medium can be estimated by evaluating the conditions of infection. Generally, the maximum titer is around 10^8 plaque-forming units (PFU) per milliliter in the medium. The lowest titer observed, for example, in the medium of the initially infected cells after several washes, is 10^3 to 10^4 PFU/ml.

Viral titers can be also estimated as $TCID_{50}$ using a 96-well plate.

IV. TRANSFER VECTORS OF *B. MORI* NUCLEAR POLYHEDROSIS VIRUS

A. Construction of Transfer Vectors

As the genome size of BmNPV is very large (about 130 kb), the only way to obtain recombinant viruses containing foreign genes is by recombination in host cells cotransfected with wild-type viral DNA and a transfer plasmid containing a foreign gene.[5] Many transfer vectors have been constructed from the plasmid pBmE36,[5] which consists of a 10.5-kb fragment containing the polyhedrin gene in pBR322. The polyhedrin gene is removed by Bal31 exonuclease[21] or by generating a recognition site using a direct mutagenesis technique.[22] Then a synthesized linker[23] or a plasmid-derived polylinker[5] is inserted after the promoter of the polyhedrin gene. These vectors generally possess an appropriate length (about 3 kb) of each of the 5' and 3' fragments for homologous recombination, a linker between these fragments for insertion of foreign genes, an antibiotic (ampicillin) resistance gene, and replication origin from a bacterial plasmid. A physical map of an example is shown in Figure 2.

B. Transfer Vectors for Expression of Mature Genes

We had already constructed a transfer vector, p89B310, which lacked 18 base pairs from the translational start in the 5' flanking sequence of polyhedrin.[5] As the 5' flanking sequence is found to be important for high levels of expression of a foreign gene,[23] a new series of transfer vectors containing almost complete 5' flanking sequence has been constructed (Figure 3).

Transfer vectors — pBK273, pBE274, pBK283, and pBE284 — possess about 3-kb fragments before and after the polyhedrin gene, as does p89B310. Unlike p89B310, these vectors contain the complete 5' flanking region except for two bases before the translational start. pBE274 has a single cloning site, *Eco*RI; pBK273 has additional sites, *Kpn*I and *Sac*I. pBE283 has a polylinker containing *Eco*RI, *Xba*I, and *Aat*I sites, and pBK284 has additional *Kpn*I and *Sac*I sites. pBE520 possesses *Eco*RV, *Xba*I, and *Aat*I sites.[24] pBM020 has a synthesized polylinker containing *Bgl*II, *Sac*I, *Sma*I, *Eco*RI, *Nco*I, *Eco*RV, *Xba*I, and *Aat*I sites (Figure 3).[23]

C. Transfer Vectors for Expression of Polyhedrin-Fused Genes

Transfer vectors of the pBF series, which are designed for making fusion proteins with the N terminus of polyhedrin, contain various lengths (4 to 133 bases from the translational start ATG) of DNA of the N-terminal coding region of the polyhedrin gene (Figure 4). It is not so difficult to presume the existence of translated polypeptides from DNA sequence data. However, it is not so easy to express an unknown protein, especially if one lacks a

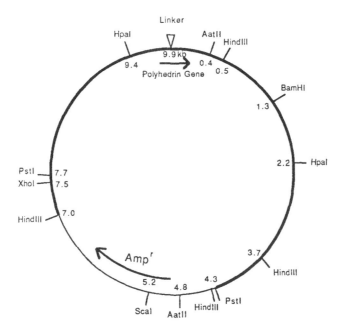

FIGURE 2.   A physical map of the transfer vector, pBE274. The start
site of the map is the multiple cloning site. The thin line indicates the
fragment derived from pUC9.

good detection system (e.g., antibody detection etc.). The pBF series of expression vectors
are constructed for the expression of this kind of polypeptide by producing fusion proteins
connected after the N terminus of the polyhedrin gene. These kinds of vectors are useful
for the expression of polypeptides not translated from the ATG or for expressing only an
interesting part of the polypeptide. Furthermore, fusion proteins containing part of the
polyhedrin protein plus part of a foreign gene are found to be expressed higher, in some
cases, than those produced as mature gene products.

p89BX40 is constructed for making fusion proteins connected at the *Xba*I site, normally
found 156 bases from the translational start of the polyhedrin gene (Figure 4).[25] A series of
12 pBF vectors are constructed by Bal31 exonuclease digestion from the end of the polyhedrin
gene, ligation to the *Sma*I site located just after the *Eco*RI site of the polylinker of pUC9,
and then ligated to the 3-kb downstream fragment of pBE284.

D. Construction of Recombinant Plasmids

The transfer vector plasmids described above contain unique recognition site(s) for insertion of foreign genes. One of the sites of the linker is chosen for ligation to the inserted
gene. If an appropriate site for insertion is not found, blunt-end ligation can be used.
Construction is easier if a foreign gene can be taken out at two different recognition sites.
For example, if the gene is removed by digestion with *Eco*RI at a site before the translational
start and another enzyme at a site after the stop codon, the site near the stop codon can be
cleaved first and treated with Klenow fragment to make a blunt end, if necessary, and then
the other site can be cleaved with *Eco*RI. This fragment can be inserted between the *Eco*RI
and *Aat*I sites in pBE384. In addition, *Sca*I digestion in the ampicillin-resistant gene of the
plasmid containing a foreign gene and phosphatase treatment of the transfer vector can
increase the efficiency of obtaining a recombinant plasmid.

Recombinant plasmids are identified by digestion with restriction enzymes or hybridization
using a foreign gene DNA as a probe. The correct orientation is confirmed by double digestion

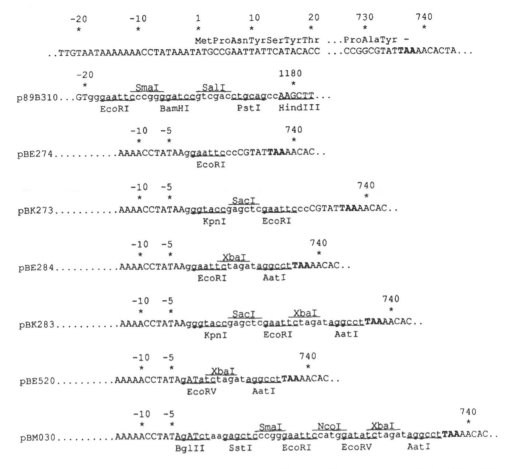

FIGURE 3. Transfer vectors for expression of mature genes. The nucleotide sequence around the translation start codon of the polyhedrin gene is shown at the top with the amino-terminal "TAA" numbered as + 1. The nucleotide of the translational stop signal, "TAA", is indicated in bold characters. The altered sequences are indicated by lower case characters.

with restriction enzymes. Sequencing using a primer corresponding to the 5' flanking region of the polyhedrin gene is also useful.

V. CONSTRUCTION OF RECOMBINANT *B. MORI* NUCLEAR POLYHEDROSIS VIRUS

A. Preparation of Viral Particles for Extraction of DNA

Viral DNA for construction of a recombinant virus is prepared from either infected cell culture fluid or infected larvae.

Preparation of virus from cultured cells is as follows. Confluent cell monolayers in 150-mm dishes (5×10^7 cells in 30 ml medium) are prepared for infection. The medium is removed by aspiration, and 500 µl to 1 ml of diluted viral solution containing about 2×10^8 PFU of wild-type BmNPV is added into each of the 150-mm dishes. After incubation for 60 min with rocking at 15-min intervals, 17 ml of the medium is added. The cell fluids are collected at 60 to 72 h postinfection and centrifuged at 3000 rpm for 20 min at 5°C. After an additional centrifugation at 3000 rpm for 20 min, the low-speed supernatants are centrifuged at $100,000 \times g$ for 60 min. The pellets are suspended in about 2 ml of 10 mM Tris-HCl (pH 7.5) and 1 mM EDTA by strong pipetting and vortexing. Routinely, 6 ml of the viral suspension is loaded onto a 20 to 50% (w/w) sucrose linear gradient prepared in

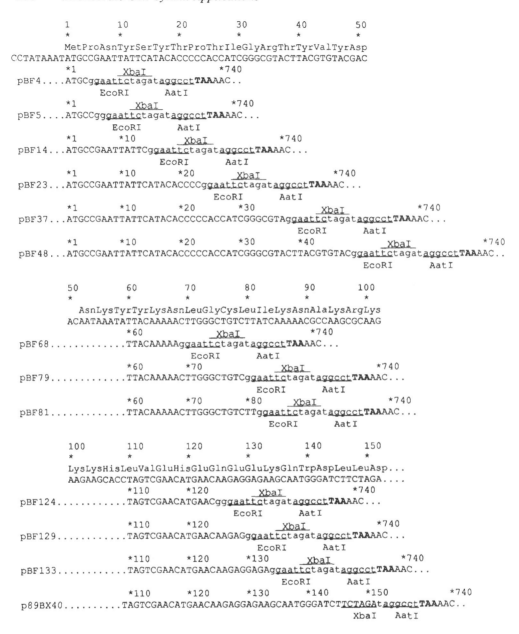

FIGURE 4. Transfer vectors for expression of polyhedrin fused genes. Twelve transfer vectors having the same polylinker are divided into three types by reading frame: type I (pBF4, pBF37, pBF79, and pBF133), type II (pBF5, pBF48, pBF81, and pBF129), and type III (pBF14, pBF23, pBF68, and pBF124). Within each type, fusion proteins are produced in the same reading frame. See the legend in Figure 3.

a 40-ml tube for a SW28 rotor. After centrifugation at 20,000 rpm for 90 min at 15°C, a viral band should be visible about $^2/_3$ of the distance down the gradient. One to two milliliters of the gradient is collected with a syringe and needle by puncturing the side of the tube. The viral solution is diluted more than 20-fold with distilled water and precipitated by centrifugation at 100,000 × *g* for 60 min. The supernatant is discarded, and the solution remaining on the surface of the tube is removed with a Kimwipe. The precipitates are suspended in a small amount of 10 m*M* Tris-HCl (pH7.5) and 1 m*M* EDTA, and 200 μl each of the solution (0.3 to 0.5 mg DNA per milliliter) is transferred into a 1.5-ml polypropylene test tube (which can be stored at $-80°C$). Average yield is about 5 μg of viral DNA from culture fluid of one 150-mm dish.

For viral DNA purification from larvae, the polyhedral inclusion bodies are first purified. A large quantity of the viral particles is purified easily as follows. Larvae of the silkworm at the first day of the fifth instar are used. About 20 μl of viral solution containing 10^5 to 10^6 PFU and usually an appropriate amount of an antibiotic (1:20 volume of 60 mg/ml of kanamycin) is injected into hemolymph using a syringe.[5] After 4 to 5 d postinfection, the larvae are collected in a bottle filled with distilled water and stored at 5°C for 1 to 10 d. Polyhedra are purified from larvae by low-speed centrifugation and can be stored at 5°C. About 10^{10} polyhedra (several hundred microliters as a packed volume) are precipitated and are suspended in about 4 ml of 0.1 M Na_2CO_3 and 0.05 M NaCl.[26] After incubation for 30 min on ice, the solution will turn a clear opal color. It is then centrifuged at 3000 rpm for 5 min, and the supernatant is layered on a 20 to 50% sucrose gradient as described above. After centrifugation, virus particles from the gradient are purified by the same method as for those from culture fluid.

B. Preparation of Viral DNA for Cotransfection

Viral DNA for cotransfection is prepared from viral particles purified as described above. Proteins are tightly bound to DNA of NPV; thus, the treatment described below is necessary for extraction of DNA. Furthermore, considering the large size of viral DNA (about 130 kb), DNA should neither be vortexed nor ethanol precipitated throughout extraction. Usually, 200 μl of the viral solution containing at least 0.3 mg DNA per milliliter is used for extraction of DNA. Ten microliters of 10% SDS and 20 μl of 20 mg/ml of Protease K (Merck) are added to the DNA in an Eppendorf tube and mixed gently by flicking with a finger. After incubation at 37°C for 2 to 5 h, the solution is extracted with an equal volume of phenol two to three times, extracted with an equal volume of phenol/chloroform (1:1), and extracted with an equal volume of chloroform twice. Each extraction is done by inverting the tube very slowly. Except for the final extraction with chloroform, the organic bottom layer is removed with a pipetteman. The purified DNA solution, which contains about 1% SDS, can be stored at least 4 months at 5°C. This concentration of SDS does not cause effects on cotransfection of BmN cells. Furthermore, digestions with most restriction enzymes are not inhibited by SDS if the solution is diluted 1:5.

C. Cotransfection of a Recombinant Plasmid and Wild-Type Viral DNA

Recombinant viruses are constructed by homologous recombination between recombinant plasmid DNA and wild-type viral DNA in calcium-mediated cotransfected cells. The method for cotransfection[5] is similar to that used in mammalian cells.[27]

Two different solutions are prepared in Eppendorf tubes. Solution A (250 μl) is composed of 2 μg of viral DNA, 5 μg of recombinant plasmid DNA in 0.25 M $CaCl_2$. Usually, 5 μl of viral DNA, 5 μl of the recombinant plasmid DNA, 30 μl of 2 M $CaCl_2$, and 210 μl of distilled water are mixed. Solution B (250 μl) is composed of 50 mM Hepes buffer (pH 7.1, 0.28 M NaCl, 0.7 mM Na_2PO_4, and 0.7 mM $NaHPO_4$). Usually, 250 μl of 50 mM Hepes and 0.28 M NaCl in a tube is added to 5 μl each of 35 mM Na_2PO_4 and 35 mM $NaHPO_4$. To make the calcium precipitate, solution A is added dropwise with a pipetteman into solution B which is gently bubbling with air from a Pasteur pipette. The color of the solution changes to slight opal upon precipitation. After incubation for 30 min at room temperature, 500 μl of the mixture is added into the culture of BmN cells in a 60-mm dish. Stock solutions of antibiotics (for example, 10 μl of 60 mg/ml kanamycin) are added to prevent contamination problems. Six to sixteen hours after adding the cotransfection mixture, the medium is replaced with fresh medium and the cells are cultured for 5 d more. Polyhedral inclusion bodies can be seen 4 to 5 d after transfection. The cell culture fluid is subjected to plaque assay for isolation of a recombinant virus.

D. Isolation of a Recombinant Virus Carrying a Foreign Gene

Recombinant virus in which the polyhedrin gene is replaced with a foreign gene is detected by the absence of production of inclusion bodies in infected nuclei using a light microscope. To isolate the recombinant virus from wild-type viruses, the plaque assay described above is used. The cell monolayer infected with a cotransfected virus sample is overlaid by 0.75% SeaPlaque agarose and cultured for 4 to 6 d. Generally, between 0.2 to 3% of the produced plaques are occlusion negative.[17]

The culture fluid at 5 to 6 d after cotransfection usually contains about 10^7 PFU/ml. The fluid is collected and diluted for plaque assay. Routinely, the following three sets of solutions are prepared by dilution with TC-100 containing 1% FCS: (A) 10^{-3} dilution — 10 μl of the transfected cell fluid is mixed into 1 ml of medium; (B) 10^{-4} dilution — 100 μl of solution (A) is added to 900 μl of medium; (C) 10^{-5} dilution — 100 μl of solution (B) is added to 900 μl of medium. One hundred (or 200) μl of each solution is added to BmN cells in a 60-mm dish for plaque assay.

Plaques produced by a recombinant virus are screened by a light microscope using a combination of low ($\times 400$) and high ($\times 1200$) magnification. Staining with neutral red on the second overlay is omitted to ease the identification of recombinant viruses by light microscopy. Five days after overlay, plaques can be easily identified without a light microscope. When dishes are observed at low magnification, the wild-type plaques show a dark color. Recombinant plaques have a slightly lighter color. It is sometimes easier to find recombinant plaques by screening with the microscope slightly out of focus. When candidates are found, they are confirmed at high magnification by the following characteristics. Polyhedral inclusion bodies are not observed, but some cells produce crystals similar to polyhedral inclusion bodies (but without clear outlines). Areas which are not completely covered with cells are typically plaques because of the inhibition of cell growth. Sometimes cells with expanded nuclei, cells with rounded shape with a distinct cell membrane, or shrunken cells are observed in the plaque.

The plaques produced by a recombinant virus are marked with a felt-tipped pen and are picked with a Pasteur pipette. The Pasteur pipette is stuck vertically into the plaque, and the agarose plug is forcefully pipetted into 1 ml of TC-100 containing 1% FCS to be broken down. This solution usually contains about 10^3 PFU. A second and third plaque purification are necessary to isolate a pure clone.

Isolation of a recombinant virus from the cotransfected sample can be also done on a 96-well plate after infection with appropriate dilutions. Cells in a well showing cytopathic effect, but not polyhedral production, indicate recombinant viral infection, and the pure isolate is cloned by an additional plating of a diluted sample from the recombinant well.

VI. EXPRESSION OF FOREIGN GENES IN INSECT CELLS

A. Characteristics of Recombinant Viruses

Hybridization and restriction endonuclease analyses show that almost all isolated viruses which do not produce polyhedral inclusion bodies are recombinant viruses with foreign genes at the expected sites. There is one example, in the case of the E2 gene of bovine papilloma virus, which shows different DNA restriction patterns from wild type, except around the polyhedrin gene. Four isolates cloned at the same time from the same plate show the same pattern as that of wild type except for the polyhedrin region. When the efficiency of production of the E2 protein is tested, no difference in production levels is detected between them.[17]

Growth curves in BmN cells of the recombinant viruses with the insertion of human α-interferon or the E2 protein of bovine papilloma virus type 2 are almost identical with that of the wild-type virus.[17] Heat inactivation curves for the two recombinant viruses at 40, 50, and 60°C showed similar patterns to those of the wild-type virus.[17]

BmN cells showed the typical cytopathic effect observed in the wild-type infection except for the lack of the production of polyhedral inclusion bodies in the nuclei.

B. Expression of Foreign Genes in BmN Cells

We have already made more than 300 different constructs of recombinant viruses with about 30 different genes from prokaryotes and eukaryotes and examined the characteristics of the polypeptides produced.[4,5,17,23-25,28] The most advantageous characteristics of this expression system *in vitro* are that the products are biologically active proteins made in their native forms at a high level.

C. Production of Secreted Proteins

Secreted proteins are first synthesized as molecules having hydrophobic signal peptides at their N termini, and are then cleaved by a peptidase on the membrane of the endoplasmic reticulum.[29]

To examine whether signal peptides of mammals were recognized and cleaved at the correct sites, the mammalian secreted gene for human α-interferon was introduced into BmNPV. The N-terminal amino acid sequence of the interferon made in BmN cells is identical with that produced in the original mammalian cells.[5]

To confirm the importance of the signal peptides for secretion, a recombinant virus without a signal sequence is constructed. That is, the ATG for the translational start is inserted before the mature interferon sequence and the sequence corresponding to the signal peptide is completely removed. During the infection with this virus lacking a signal peptide sequence, the interferon activity is found mainly within the BmN cells.[4]

Furthermore, mouse interleukin 3 expressed in BmN cells is found to be glycosylated. This secreted protein is also secreted into the medium, and the N terminus of the interleukin is identical with that produced in a mammalian cell line.[24]

D. Production of Polyhedrin Fusion Proteins

The E2 protein of bovine papilloma virus has not been detected in infected tissues; however, the gene has been cloned into a plasmid, and the translation product is deduced.[30] For studies of this kind of unknown protein, we constructed a recombinant virus which can produce a fusion protein with the polyhedral inclusion body protein. This fusion protein is separated on SDS polyacrylamide gel and is detected by antibody prepared to the polyhedral protein. In this case, this fusion protein is easily identified by its abundant major band on the gel. Due to the strong promoter of the polyhedrin gene, this E2 fusion protein is synthesized to levels of more than 10% of the total protein.[25]

E. Expression of Foreign Genes in the Larvae of the Silkworm

The other unique characteristic of the BmNPV expression system is the ease of expansion of the *in vitro* system to an *in vivo* system using larvae. If the productivity is compared between BmN cells and larvae, the value for the larvae is usually higher than that for BmN cells.[5,24]

With respect to safety considerations of recombinant DNA experiments, the host silkworm and BmNPV are a very well-suited system. Silkworms lack the ability to live outside, and they are not found in the field. BmNPV has a very narrow host range, almost limited to one species. Furthermore, recombinant viruses which do not contain a polyhedrin gene are easily inactivated and show very low infectivity by per os infection, the only natural means of transmission of virus.

VII. CONCLUSION

As described above, the BmNPV expression system has many advantageous characteristics for expression of foreign genes. I hope that this system will make many contributions to basic biology and also to applied science, such as production of pharmaceuticals, in the near future.

ACKNOWLEDGMENTS

These experiments were performed in collaboration with several scientists. I would like to thank Drs. M. Obinata, Y. Saeki, T. Horiuchi, Y. Sato, and M. Furusawa for their collaboration on the expression of human α-interferon, Drs. A. Fuse and B. Simizu for their collaboration on the expression of the E2 protein of bovine papilloma virus, and Drs. A. Miyajima and K. Arai for their collaboration on the expression of mouse interleukin 3. I would also like to thank Dr. A. H. Roter for critical reading of the manuscript.

REFERENCES

1. **Matthews, R. E. F.**, Classification and nomenclature of viruses, *Intervirology*, 17, 1, 1982.
2. **Smith, G. E., Summers, M. D., and Fraser, M.**, Production of human β interferon in insect cells infected with a baculovirus expression vector, *Mol. Cell. Biol.*, 3, 2156, 1983.
3. **Pennock, G. D., Shoemaker, C., and Miller, L. K.**, Strong and regulated expression of Escherichia coli β-galactosidase in insect cells with a baculovirus vector, *Mol. Cell. Biol.*, 4, 394, 1984.
4. **Maeda, S., Kawai, T., Obinata, M., Chika, T., Horiuchi, T., Maekawa, K., Nakasuji, K., Saeki, Y., Sato, Y., Yamada, K., and Furusawa, M.**, Characterization of human interferon-α produced by a gene transferred by a baculovirus vector in the silkworm, *Bombyx mori*, *Proc. Jpn. Acad. Ser.*, B60, 423, 1984.
5. **Maeda, S., Kawai, T., Fujiwara, H., Horiuchi, T., Saeki, Y., and Furusawa, M.**, Production of human α-interferon in silkworm, using a baculovirus vector, *Nature (London)*, 315, 592, 1985.
6. **Maeda, S.**, *Annu. Rev. Entomol.*, 34, 351, 1989.
7. **Maeda, S.**, Expression of human interferon in silkworms with a baculovirus vector, in *Invertebrate Pathology and Cell Culture*, Maramorosch, K., Ed., Academic Press, New York, in press.
8. **Rohrman, G. F.**, Polyhedrin structure, *J. Gen. Virol.*, 67, 1499, 1986.
9. **Grace, T. D. C.**, Establishment of four strains of cells from insect tissues grown in vitro, *Nature (London)*, 195, 788, 1962.
10. **Wood, H. A.**, An agar overlay plaque assay method for *Autographa californica* nuclear-polyhedrosis virus, *J. Invertebr. Pathol.*, 29, 304, 1977.
11. **Volkman, L. E. and Goldsmith, P. A.**, Generalized immunoassay for *Autographa californica* nuclear polyhedrosis virus infectivity in vitro, *Appl. Environ. Microbiol.*, 44, 227, 1982.
12. **Maeda, S.**, A plaque assay and cloning of *Bombyx mori* nuclear polyhedrosis virus, *J. Seric. Sci. Jpn.*, 53, 547, 1984.
13. **Raghow, R. and Grace, T. D. C.**, Studies on a nuclear polyhedrosis virus in *Bombyx mori* cells in vitro. I. Multiplication kinetics and ultrastructural studies, *J. Ultrastruct. Res.*, 47, 384, 1974.
14. **Inoue, H. and Mitsuhashi, J.**, A *Bombyx mori* cell line susceptible to a nuclear polyhedrosis virus, *J. Seric. Sci. Jpn.*, 53, 108, 1984.
15. **Ninaki, O.**, personal communication.
16. **Quiot, J. M.**, Establishment of a cell line from the ovaries of *Bombyx mori*, L. (Lepidoptera), *Sericologia*, 22, 25, 1982.
17. **Maeda, S.**, unpublished data, 1984—1987.
18. **Maekawa, H. and Tamura, T.**, personal communication, 1986.
19. **Gardiner, G. R. and Stockdale, H.**, Two tissue culture media for production of lepidopteran cells and nuclear polyhedrosis viruses, *J. Invertebr. Pathol.*, 25, 363, 1975.
20. **Hink, W. F. and Vail, P. V.**, A plaque assay for titration of alfalfa looper nuclear polyhedrosis virus in a cabbage looper (TN-368) cell line, *J. Invertebr. Pathol.*, 22, 168, 1973.
21. **Guo, L.-H. and Wu, R.**, Exonuclease. III. Use for DNA sequence analysis and in specific deletions of nucleotides, in *Methods in Enzymology*, Vol. 100, Wu, R., Grossman, L., and Moldave, K., Eds., 1983, 60.
22. **Razin, A., Hirose, T., Itakura, K., and Riggs, A. D.**, Efficient correction of a mutation by use of chemically synthesized DNA, *Proc. Natl. Acad. Sci. U.S.A.*, 75, 4268, 1978.

23. **Horiuchi, T., Marumoto, Y., Saeki, Y., Sato, Y., Furusawa, M., Kondo, A., and Maeda, S.,** High-level expression of the human-α-interferon gene through the use of an improved baculovirus vector in the silkworm, *Bombyx mori, Agric. Biol. Chem.,* 51, 1573, 1987.

24. **Miyajima, A., Schreurs, J., Otsu, K., Kondo, A., Arai, K., and Maeda, S.,** Use of the silkworm, *Bombyx mori,* and an insect baculovirus vector for high-level of expression and secretion of biologically active mouse interleukin-3, *Gene,* 58, 273, 1987.

25. **Tada, A., Fuse, A., Sekine, H., Simizu, B., Kondo, A., and Maeda, S.,** Expression of E2 open reading frame of papillomaviruses BPV1 and HPV6b in silkworm by a baculovirus vector, *Virus Res.,* 9, 375, 1988.

26. **Kawarabata, T. and Matsumoto, K.,** Isolation and structure of a nuclear polyhedrosis virus from polyhedra of the silkworm, *Bombyx mori, Appl. Entomol. Zool.,* 8, 227, 1973.

27. **Graham, F. L. and Van der Eb, A. J.,** A new technique for the assay of infectivity of human adenovirus 5 DNA, *Virology,* 52, 456, 1973.

28. **Marumoto, Y., Sato, Y., Fujiwara, H., Sakano, K., Saeki, Y., Agata, M., Furusawa, M., and Maeda, S.,** Hyperproduction of polyhedrin-IGF II fusion protein in silkworm larvae infected with recombinant *Bombyx* nuclear polyhedrosis virus, *J. Gen. Virol.,* 68, 2599, 1987.

29. **Wickner, W. T. and Lodish, H. F.,** Multiple mechanisms of protein insertion into and across membranes, *Science,* 230, 400, 1985.

30. **Chen, E. Y., Howley, P. M., Levinson, A. D., and Seeberg, P. H.,** The primary structure and genetic organization of the bovine papilloma virus type 1 genome, *Nature (London),* 299, 529, 1982.

Chapter 20

EFFICIENT EXPRESSION OF FOREIGN GENES IN CULTURED *DROSOPHILA MELANOGASTER* CELLS USING HYGROMYCIN B SELECTION

Ariane van der Straten, Hanne Johansen, Raymond Sweet, and Martin Rosenberg

TABLE OF CONTENTS

I. INTRODUCTION

Recombinant DNA technology has made the efficient expression of foreign gene products possible in a variety of heterologous cell systems. The bacteria, led by *Escherichia coli*, have been the most exploited hosts for gene expression. However, certain limitations of these systems, particularly in the expression of posttranscriptionally modified eukaryotic proteins, has led to the development of several eukaryotic host cell expression systems. A relatively recent addition to this collection of hosts is cultured insect cells.[1] These cells have several attractive growth features including their ability to grow in suspension at room temperature, under atmospheric conditions, and to a high cell density (1 to 2 × 10[7] cells per milliliter). Also, they offer a more sophisticated eukaryotic cell environment than does yeast for studying the expression of functional mammalian gene products. It is known that insect cells carry out many of the typical higher eukaryotic posttranslational modification events such as those needed for secretion, glycosylation, or subcellular localization.[3-5]

To date, expression of heterologous genes in insect cells has been restricted primarily to the lytic baculovirus system.[3,4,6] However, recently the stable cotransfection of high copy numbers of *Drosophila* genes into the *Drosophila melanogaster* Schneider S2 cell line and the expression at high levels of their corresponding mRNA was described.[2,7] This observation led us to explore transfection of S2 cells as a method for the overexpression of heterologous gene products. A few systems for dominant selection of transfected genes in *Drosophila* cells have been described.[8-11] Unfortunately, they suffer from several disadvantages. For example, methotrexate selection, while efficient, requires as long as 2 months for the outgrowth of resistant cells. G418 selection, although more rapid, yields a high background of nontransformed resistant cells. Therefore, we sought to develop a rapid and efficient system for transfection and expression of foreign genes into the *D. melanogaster* S2 cell line.

We describe here a novel hygromycin B (hygro) selection system which yields stable transformants within 3 weeks with no background of spontaneous resistant cells. Using the *E. coli* gene for the galactokinase (*galK*) enzyme, we demonstrate the stable cotransfection of a nonselected gene at high copy number and the high level expression of its functional product. We also compare the relative efficiencies of three promoters: the *Drosophila* COPIA 5′LTR, the *Drosophila* metallothionein (MT) promoter, and the mammalian viral SV40 early promoter. Finally, we demonstrate the utility of the metallothionein promoter to regulate the expression of the human cH-*ras* gene and its oncogenic val[12] mutant which are toxic when overexpressed in S2 cells.

II. RESULTS

A. Selection Systems for Introduction of Foreign Genes into Insect Cells

1. Vector Construction

To compare selection with hygro, methotrexate (mtx), and G418, genes encoding resistance to these drugs were placed under the transcriptional control of the COPIA 5′LTR and flanked downstream by the SV40 early polyadenylation signal.[12,13] The vector pHGCO (Figure 1), kindly provided by Peter Cherbas, carries a gene coding for the bacterial dihydrofolate reductase (dhfr) which is resistant to mtx.[8,14] The vectors pCO*neo and pCOdhygro carry the neomycin phosphotransferase gene from Tn5, or the *E. coli* hygro phosphotransferase gene, which confer resistance to the antibiotics G418 and hygro, respectively (Figure 1).[15-18]

2. Comparison of the Selection Systems

S2 cells were transfected by Ca-PO₄ coprecipitation with the vectors carrying the resistance genes, and then grown and selected in the presence of the appropriate drug.[19,20] Over a

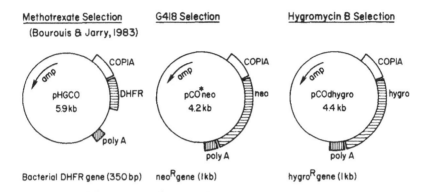

FIGURE 1. Vectors for expression of the resistance genes. The pHGCO vector has been described.[8] pCO*neo was constructed by inserting a ~1-kb *Bgl*II-*Sma*I fragment carrying the neo[R] gene from Tn5 into pUCOPIA, followed by the SV40 early poly A region deriving from DSPI.[13,15] pUCOPIA is a pUC18 plasmid with a ~350-bp *Bam*HI fragment containing the copia 5'LTR from pHGCO. pCOdhygro was constructed by cloning into a pUCOPIA derivative a ~1.48-kb *Hind*III-*Bam*HI fragment carrying the hygro[R] gene followed by the SV40 early poly A from DSPhygro.[35]

Table 1
COMPARISON OF SELECTION SYSTEMS

Drug	Concentration (per ml)	Vector	Selection time (weeks)	Background resistance
Methotrexate	0.1 μg	pHGCO	6	0
G418	1 mg	pCO*neo	3	+
Hygromycin B	300 μg	pCOdhygro	2—3	0

period of 2 to 6 weeks, cell viability declined and then recovered with the outgrowth of resistant cells. To avoid clonal bias, all our experiments were carried out with these polyclonal populations. As summarized in Table 1, selection with hygro results in a stable, resistant cell population within 2 to 3 weeks. This is twice as rapid as observed for mtx. Moreover, hygro selection does not give rise to the high background of spontaneous resistance observed with G418 selection.

B. Cointroduction and Expression of the *E. coli galK* Gene in S2 Cells

We have utilized the hygro selection system to stably cointroduce and examine expression of a second non-*Drosophila* gene in the S2 cells. To this end, the gene for the *E. coli* enzyme *galK* was inserted into an expression cassette containing the control elements for initiation of transcription (COPIA 5'LTR) and polyadenylation (SV40 early poly A) to give the vector pCOdGALK. As outlined in Figure 2, the vectors pCOdGALK and pCOdhygro were cotransfected into S2 cells at different ratios of DNA (total transfected DNA was always 20 μg), and cells were selected for resistance to hygro. The resultant stable transfectants were probed for DNA copy number and *galK* RNA and protein levels.

1. Gene Copy Number Correlates with the Relative Amount of Vector DNA in the Transfection

High molecular weight (MW) DNA was extracted from the resistant cell populations and probed for the transfected *galK* and hygro coding sequences by Southern blotting analysis.[21] The results showed that both genes are stably integrated in the host cellular DNA (Figure 3). The copy number of the *galK* gene in the different cell lines varied from 40 to 1050

Overexpression of E.coli Galactokinase in S2 cells

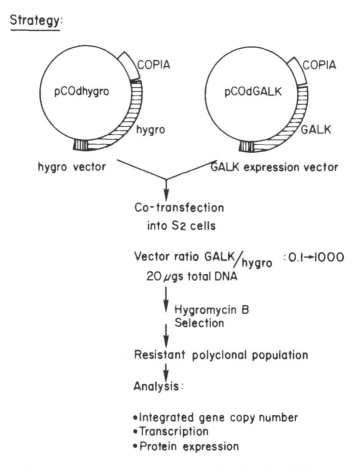

FIGURE 2. Strategy of cotransfection into S2 cells. pCOdGALK and pCOdhygro were cotransfected into S2 cells at ratios varying from 0.1 to 1000. A total of 20 μg of DNA was used in each transfection. The S2 cell line was grown in M3 medium supplemented with 12% heat-inactivated fetal bovine serum.[36] For selection and maintenance of the resistant cell populations, 300 μg/ml of hygromycin B was added to the medium.

and was approximately proportional to the amount of pCOdGALK relative to pCOdhygro in the DNA transfection experiment (Figure 3 and Table 2). Similarly, the copy number of the hygro gene varied from 400 to 1 to 2 with decreasing amounts of pCOdhygro in the transfection. These results show that hygro selection can be used to introduce the bacterial *gal*K gene at high copy number into S2 cells and that one or two copies per cell of the hygro gene are sufficient to confer resistance to the drug.

2. Comparison of Promoter Signals

The *gal*K gene was placed under the control of three different promoter signals, and expression in S2 cells was assessed. The resulting expression vectors, pCOdGALK, pDMTGALK, and SV40GALK, harbor the constitutive promoter from the COPIA 5′LTR, the inducible *Drosophila* MT promoter, and the simian virus SV40 early promoter, respectively (Figure 4).[12,13,22] These vectors were introduced into S2 cells by cotransfection with pCOdhygro and selection with hygro.

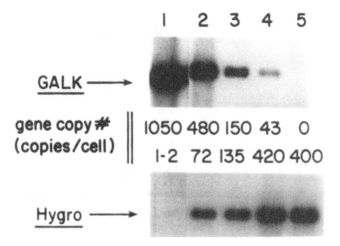

FIGURE 3. *Gal*K gene copy number in transfected S2 cells. Southern blot analysis was performed on DNA from cultures cotransfected with the following amounts of pCOdGALK and pCOdhygro vectors: 20 to 0.02 μg (lane 1), 18 to 2 μg (lane 2; cell line C), 10 to 10 μg (lane 3; cell line B), 2 to 18 μg (lane 4; cell line A), and 20 μg of pCOdhygro alone (lane 5; control line D). High MW DNA was extracted from the stable transfectants and digested with *Nco*I and *Eco*RI restriction endonucleases which cut at sites bounding the *gal*K and hygro genes; 250 ng of each digestion was run in duplicate on a 0.8% agarose gel and transferred onto nitrocellulose filters. The presence of stably integrated genes was detected by hybridization of the filters with ^{32}P-nick-translated *gal*K- and hygro-DNA probes followed by autoradiography of the filters.[21] Copy number was deduced by comparison to known amounts of digested vector DNA.

Table 2
EXPRESSION OF *E. COLI galK* IN *DROSOPHILA* CELLS

Promoter		DNA gene copy no. (copies/cell)	RNA Level[a] (pg/μg)	RNA/DNA[b] (× 10)	Protein level (ng/μg)
Copia	A	43	4	≈1	0.9
	B	150	6	0.4	1.3
	C	480	8	0.16	2.3
MT	F	210 (−)Cd	≈1	0.047	0.03
		(+)Cd	44	2.0	3.0
SV40 (early)	E	87	<0.6	—	—

[a] *gal*K mRNA levels were measured by dot blot analysis of total cellular RNAs and compared to known amounts of digested plasmid DNA, using a ^{32}P-nick-translated *gal*K DNA probe.
[b] RNA/DNA ratio normalizes the amount of *gal*K RNA per gene copy in each of the cell cultures.

The expression of *gal*K RNA and protein was measured in each of the resultant cell lines. By RNA blot analysis, both the COPIA and MT vectors gave *gal*K transcripts of the expected size, 1.5 kb, and the RNA levels increased with increasing gene copy number (Figure 5, lane A → C). Cells transfected with the MT vector were grown in the presence of cadmium to fully induce the MT promoter (Figure 5, lane F(+), see below). In contrast, only very low levels of *gal*K RNA were observed with the SV40 promoter (Figure 5, lane E).

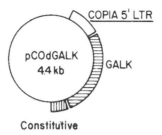

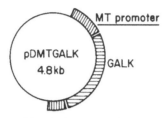

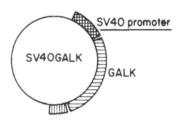

FIGURE 4. *galK* expression vectors. pCOdGALK is a pU-COPIA derivative which carries a 1.38-kb *Hind*III-*Bam*HI fragment from DSPI encoding the *E. coli galK* gene followed by the SV40 early poly A.[35] pDMTGALK has a 430-bp *Eco*RI-*Stu*I fragment coding for the *Drosophila* MT promoter followed by the same *Hind*III-*Bam*HI *galK* cassette 3.[37] SV40GALK is a derivative of the DSP1 vector.[38]

The expression of *galK* protein was monitored by immunoblot analysis (Figure 6).[23] The recombinant *galK* product was of the expected size, 41 kDa, indicating that *Drosophila* cells appropriately express the bacterial gene. In general, the amount of *galK* product in the different cell lines reflected the mRNA levels and, hence, the relative strength of the promoters controlling expression (Table 2).

These experiments show that transcription and protein expression increase with *galK* gene copy, although the relative efficiency of transcription appears to decrease with increasing copy number (see Table 2, lanes A,B,C). Our measurements of steady-state RNA levels indicate that at comparable gene copy number, the MT promoter is five- to seven-fold more efficient than the COPIA LTR and more than 100 times stronger than the SV40 early promoter. However, the higher RNA levels observed with the MT promoter did not result in a proportionately higher level of the *galK* gene product.

C. Regulation of the *Drosophila* Metallothionein Promoter

Metallothioneins are small CYS-rich proteins that bind heavy metals. They accumulate in response to divalent cations such as Cu^{2+}, Cd^{2+}, or Zn^{2+}, and studies have shown that the induction occurs at the RNA level and involves *de novo* transcription.[24] Lastowski-Perry and co-workers have isolated the *D. melanogaster* MT gene and its regulatory sequences.[22]

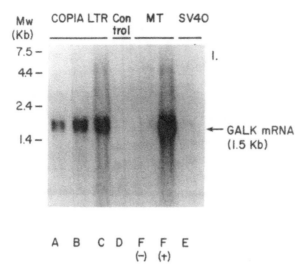

FIGURE 5.   RNA analysis of S2 cells expressing the *E. coli gal*K gene. RNA levels in S2 cells cotransfected with the different GALK expression vectors and pCOdhygro were assayed by Northern blot analysis: COPIA GALK cell lines with low (A), medium (B), and high (C) gene copy number; control cell line with no GALK (D); MT GALK cell line (F) uninduced (−) or induced (+) with 10 μ*M* CdCl$_2$ for 24h; SV 40 GALK cell line (E). Total RNA was extracted from 2 × 10⁶ cells,[39] run on a 1.2% formaldehyde-agarose gel, and transferred onto a nitrocellulose filter. Specific mRNAs were detected by hybridization of the filter with a ^{32}P-nick-translated *gal*K-DNA probe, followed by autoradiography of the filter.[21]

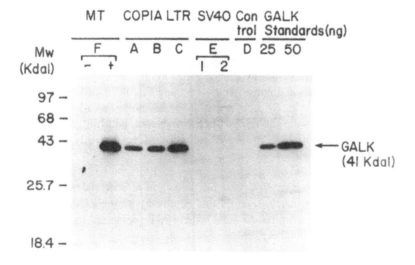

FIGURE 6.   *gal*K protein expression in transfected S2 cells. Protein samples from the cotransfected lines were assayed for *gal*K protein by Western blot analysis. Lane F: protein extracts from the MT GALK cell line uninduced (−) or induced (+) for ~24 h with 10 μ*M* CdCl$_2$; lanes A-C: protein extracts from COPIA GALK cell lines with low (A), medium (B), and high (C) gene copy number; lane E: 30 μg (1) and 60 μg (2) of protein extracts from SV40 GALK cell line; lane D: control cell line; 30 μg of proteins from total cell lysate were run on a 12%-SDS-PAGE and transferred onto a nitrocellulose filter. The filter was then treated with rabbit anti-*gal*K antiserum followed by ¹²⁵I-protein A.[23]

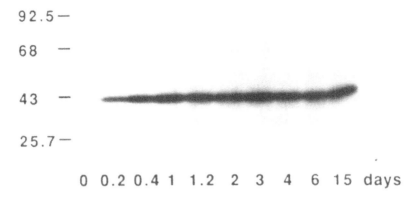

FIGURE 7.   Time course of Cd induced *gal*K protein expression. The level of *gal*K expressed under continuous metal induction in a cell line transfected with pDMTGALK was measured by Western blot analysis. Cells were grown in the presence of 10 μM of $CdCl_2$ for 2 weeks. Samples were removed at intervals and extracts from 10^5 cells were run on a 10%-SDS-PAGE and transferred onto a nitrocellulose filter. The filter was then treated as described in Figure 6.

In the previous section, we showed that the *Drosophila* MT promoter element can be used to efficiently express the *E. coli gal*K gene product in S2 cells transfected with the pDMTGALK vector. In this section, we present studies on the regulated expression of this promoter with heavy metals.

As shown in the previous section, in the presence of cadmium the cultures carrying the pDMTGALK vector expressed high levels of both *gal*K RNA and protein (Figures 5 and 6). For example, addition of 10 μM $CdCl_2$ in the culture medium results in a ~50-fold increase in both *gal*K RNA and protein (see Table 2). On the contrary, in the absence of added metal ion, only very low levels of transcript and gene product were detected. Clearly, the transfected MT promoter can be fully regulated and selectively induced by the addition of heavy metals.

We followed the induction of *gal*K expression with time and found that expression began soon after addition of Cd. The response reached a maximum after 24 h and remained constant for at least 2 weeks if Cd was maintained in the medium (Figure 7). Moreover, we compared the efficiency of cadmium ($CdCl_2$), copper ($CuSO_4$), and zinc ($ZnSo_4$) for induction of the MT promoter. Cd and Cu were the most effective inducers, and full *gal*K expression was achieved with metal concentrations of 5 to 10 and 200 μM, respectively (Figure 8). Zn was less efficient (data not shown).

It is notable that even at high gene copy number (>200), the basal expression from the *Drosophila* MT promoter is very low, and full induction is achieved with the addition of metal ion, to a level approximately fivefold that obtained with the COPIA LTR. This is in contrast to observations in mammalian cells, where MT promoters lose their inducibility when present in high copy number.[24,25]

D. Regulated Overexpression of the Human Ha-*ras* Oncogene in S2 Cells

The *ras* genes are phylogenetically among the most conserved oncogenes, and close homologs are found in organisms ranging from yeast to man.[26] These genes are implicated in mammalian cell growth regulation, and point mutations resulting in single, amino acid changes in the *ras* proteins, p21, are observed in a variety of tumor cells and tumor cell lines.[26,27] Introduction of these mutant *ras* genes into certain cell lines, such as NIH-3T3 cells, or overexpression of a normal *ras* gene results in growth transformation.[28] These transformed cells show a reduced requirement for added growth factors and serum supple-

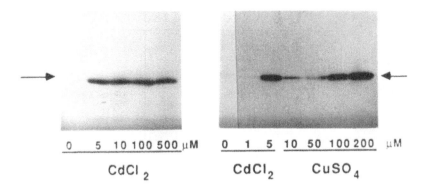

FIGURE 8. Induction of the MT promoter with Cd and Cu. The indicated concentrations of $CdCl_2$ or $CuSO_4$ were added to a pDMTGALK-transfected cell line. Samples were removed after 24 h, and total cell extracts from 10^5 cells were run on a 10%-SDS-PAGE and treated as described in Figure 6.

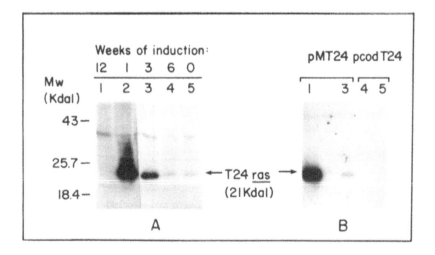

FIGURE 9. Loss of T24 *ras* expression under permanent induction of the MT promoter. (A) pMT24 transfected cells were seeded in complete medium containing 10 μM of $CdCl_2$; 30 μg of protein extracts from the cells before (lane 5) or after 1 week (lane 2), 3 weeks (lane 3), 6 weeks (lane 4), and 12 weeks (lane 1) of induction were analysed on a 15%-SDS-PAGE and transferred onto a nitrocellulose filter. The filter was subsequently treated with rabbit anti-*ras* antiserum followed by ^{125}I-protein A.[23] (B) pMT24 transfected cells were maintained in culture for 10 months, then induced for 48 h with Cd (lane 1) or uninduced (lane 3), and analyzed as in (A). Survivors from pCOdT24 transfected cultures were analysed for constitutive *ras* expression (lanes 4 and 5).

ments.[29,30] In an attempt to examine the effects of *ras* on the growth properties of insect cells, we introduced and expressed both the normal and transforming val[12] mutant human H-*ras* genes into the *Drosophila* S2 cell line. We describe here our results for the val[12] mutant *ras* oncogene (T24 *ras*).[31] Similar effects, although less pronounced, were observed with the normal *ras* gene.

We initially placed the T24 *ras* cDNA under the control of the COPIA 5′LTR (pCOdT24 vector) to achieve constitutive expression of the val[12]-p21 oncogene product in insect cells. pCOdT24 was introduced into S2 cells by cotransfection with pCOdhygro and selection for hygro resistance. High levels of *ras* protein were detected in transient assays 4 to 8 d after transfection, but none of the stable transfectants examined a month later expressed detectable levels of *ras* (Figure 9, panel B). During the selection procedure, the cultures went through

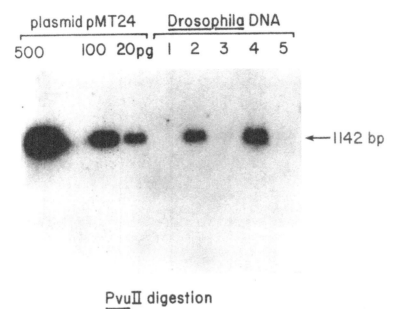

PvuⅡ digestion

FIGURE 10.   DNA analysis of *ras* transfected cells. High MW DNA was extracted from the cells before or after 2 ¹/₂ months of Cd induction. 250 ng of DNA was digested with *Pvu*II which excises the *ras* gene from the vector, run on 0.8% agarose gel, and transferred onto a nitrocellulose filter.[21] The presence of integrated genes in the host DNA was detected with a ^{32}P-nick-translated-*ras* probe. Lane 1: control; lanes 2 and 4: pMT24 and pMT*ras*wt transfected cells; lanes 3 and 5: same cultures after 2 ¹/₂ months under Cd induction.

a period of crisis resulting in viable cells which still carried the *ras* gene but had lost the ability to express the recombinant product. This effect was dependent upon the expression of the *ras* protein, because control cells stably transfected with a vector carrying a frameshift/ termination mutation in the 5' end of the *ras* gene were not affected.

To confirm our observation of apparent *ras* toxicity, we placed the T24 *ras* gene under the control of the inducible MT promoter (pMT24 vector) in order to achieve its regulated expression in S2 cells. The cells were grown in the absence of added metal ion and experienced a mild crisis which was not as pronounced nor as prolonged as that observed for cells transfected with pCOdT24. The resulting stable transfectants contained the *ras* oncogene but gave little to no expression of *ras* in the absence of added Cd. However, upon induction with Cd, cells expressed high levels of the p21 product (Figure 9). As determined by protein immunoblot analysis, the amount of *ras* protein expressed in the induced cultures was ~0.2 to 0.5% of total cellular protein. These cells remained completely inducible for *ras* expression even after 10 months in culture (Figure 9, panel B). A sample of these cultures was concomitantly maintained under permanent Cd induction for a period of ~3 months. No alteration in cell morphology or viability was detected in the Cd-induced cultures. However, a gradual loss in *ras* expression was observed over about a 6-week interval (Figure 9, panel A). Southern blot analysis of the cells before and after permanent Cd induction showed that the loss of expression resulted from the loss of the integrated *ras* DNA sequences in the induced cell population (Figure 10).

Thus, as observed for cells transfected with pCOdT24, expression of the mutant *ras* gene is toxic or at least growth static. Perhaps relevant to our observation, expression of transforming mutant *ras* genes in the murine neuronal cell line PC-12 results in differentiation and cessation of growth.[26,32] In yeast, cells expressing a chimeric human/yeast mutated *ras* gene also lose viability.[33] Recently, Bishop and Corces have introduced a *Drosophila ras* gene into the germline of the fruit fly. They observed that the expression of an "activated" version of *ras* led to dramatic effects ranging from various abnormalities to lethality, depending on the developmental stage at which the expression was induced.[34] Their results correlate well with our *in vitro* observations. The expression of the *ras* oncogene in *Drosophila* appears to be a lethal event, in contrast to its usual effect in mammalian cell systems of promoting cell proliferation and serum-free growth. We do not know the reasons for *ras* toxicity in insect cells. However, the survival of pMT24 transformants in the absence of added metal ions demonstrates the utility of the inducible MT promoter system for the regulated expression of potentially lethal gene products.

III. GENERAL CONCLUSIONS

A rapid hygro selection system has been developed for the introduction of foreign genes into the *Drosophila* S2 cell line. These cells can acquire up to a thousand copies of the transfected gene(s), and gene copy number can be modulated by varying the amount of vector DNA in the transfection. Cell lines expressing high levels of the transfected gene product(s) can be obtained within 3 weeks. Expression levels increase with gene copy number, although not in direct proportion. The recombinant products are properly localized and biochemically active. Vectors for constitutive (COPIA 5′LTR) and inducible (MT promoter) expression have been developed. The MT promoter is about fivefold more efficient than the COPIA 5′LTR, whereas the mammalian SV40 early promoter is only weakly active in *Drosophila* cells. The MT promoter is tightly regulated by cadmium and other heavy metals and remains fully inducible even at high gene copy number. RNA and protein levels are very low in the absence of metal, but both are induced 30- to 50-fold within 24 h after the addition of cadmium. High-level expression of the recombinant product is maintained when inducer is present in the cell culture.

We have observed a strong selection against transfected cells expressing the recombinant human T24 *ras* protein. Stable transfectants which constitutively expressed the *ras* oncogene could not be obtained. This phenomenon apparently results from a toxic or a growth-static effect of the *ras* protein on the cells. Notably, the regulated expression of the oncogene under the control of the MT promoter was sufficiently tight to allow the establishment and the maintenance of viable stable transformants. To our knowledge, this is the first example of the regulated expression of a lethal gene product in a higher cell system. Thus, the MT vector system will be useful for the expression of any gene with potentially detrimental effects in insect cells. Moreover, the combination of very low basal expression with efficient and rapid induction provides an attractive system for the biochemical analysis of a variety of gene products.

ACKNOWLEDGMENTS

We thank P. Cherbas for providing us with the pHGCO vector, G. Maroni for the MT promoter, and M. Reff for the DSPI and derivative vectors.

REFERENCES

1. **Fallon, A. M. and Willis, J. H.,** Gene transfer in insect cells, *Trends Biotechnol.,* 3, 217, 1985.
2. **Schneider, I.,** Cell lines derived from late embryonic stages of *Drosophila melanogaster, J. Embryol. Exp. Morphol.,* 27, 353, 1972.
3. **Miyamoto, C., Smith, G. E., Farrell-Towt, J., Chizzonite, R., Summers, M. D., and Ju, G.,** Production of human c-*myc* protein in insect cells infected with a baculovirus expression vector, *Mol. Cell. Biol.,* 5, 2860, 1985.

4. **Smith, G. E., Ju, G., Ericson, B. L., Moshera, J., Lahm, H.-W., Chizzonite, R., and Summers, M. D.,** Modification and secretion of human interleukin 2 produced in insect cells by a baculovirus expression vector[B], *Proc. Natl. Acad. Sci. U.S.A.,* 82, 8404, 1985.

5. **Allday, M. J., Sinclair, J. H., MacGillivray, A. J., and Sang, J. H.,** Efficient expression of an Epstein-Barr nuclear antigen in *Drosophila* cells transfected with Epstein-Barr virus DNA, *EMBO J.,* 4, 2955, 1985.

6. **Madisen, L., Travis, B., Hu, S.-L., and Purchio, A. F.,** Expression of the human Immunodeficiency Virus *gag* gene in insect cells, *Virology,* 158, 248, 1987.

7. **Moss, R., Cherbas, L., Koehler, M., Bunch, T., and Cherbas, P.,** Transformation of *Drosophila* cells in culture: plasmid recombination and expression of non-selectable markers, *In Vitro,* 21(3) Part II, 42A, 1985.

8. **Bourouis, M. and Jarry, B.,** Vectors containing a prokaryotic dihydrofolate reductase gene transform *Drosophila* cells to methotrexate-resistance, *EMBO J.,* 2, 1099, 1983.

9. **Rio, D. C. and Rubin, G. M.,** Transformation of cultured *Drosophila melanogaster* cells with a dominant selectable marker, *Mol. Cell. Biol.,* 5, 1833, 1985.

10. **Maisonhaute, C. and Echalier, G.,** Stable transformation of *Drosophila* Kc cells to antibiotic resistance with the bacterial neomycin resistance gene, *FEBS Lett.,* 197, 45, 1986.

11. **Sinclair, J. H., Sang, J. H., Burke, J. F., and Ish-Horowicz, D.,** Extrachromosomal replication of *copia*-based vectors in cultured *Drosophila* cells, *Nature (London),* 306, 198, 1983.

12. **Flavell, A. J., Levis, R., Simon, M. A., and Rubin, G. M.,** The 5' termini of RNAs encoded by the transposable element copia, *Nucleic Acid Res.,* 9, 6279, 1981.

13. **Pfarr, D. S., Sathe, G., and Reff, M. E.,** A highly modular cloning vector for the analysis of eukaryotic genes and gene regulatory elements, *DNA,* 4, 461, 1985.

14. **O'Hare, K., Benoist, C., and Breathnach, R.,** Transformation of mouse fibroblasts to methotrexate resistance by a recombinant plasmid expressing a prokaryotic dihydrofolate reductase, *Proc. Natl. Acad. Sci. U.S.A.,* 78, 1527, 1981.

15. **Beck, E., Ludwig, G., Auerswald, E. A., Reiss, B., and Schaller, H.,** Nucleotide sequence and exact localization of the neomycin phosphotransferase gene from transposon Tn5, *Gene,* 19, 327, 1982.

16. **Gritz, L. and Davies, J.,** Plasmid-encoded hygromycin B: the sequence of hygromycin B phosphotransferase gene and its expression in *Escherichia coli* and *Saccharomyces cerevisiae, Gene,* 25, 179, 1983.

17. **Davies, J. and Jimenez, A.,** A new selective agent for eukaryotic cloning vectors, *Am. J. Trop. Med. Hyg.,* 29, (Suppl.) 1089, 1980.

18. **Santerre, R. F., Allen, N. E., Hobbs, J. N., Jr., Rao, N., and Schmidt, R. J.,** Expression of prokaryotic genes for hygromycin B and G418 resistance as dominant-selection markers in mouse L cells, *Gene,* 30, 147, 1984.

19. **Di Nocera, P. P. and Dawid, I. B.,** Transient expression of genes introduced into cultured cells of *Drosophila, Proc. Natl. Acad. Sci. U.S.A.,* 80, 7095, 1983.

20. **Wigler, M., Sweet, R., Sim, G. K., Wold, B., Pellicer, A., Lacey, E., Maniatis, T., Silverstein, S., and Axel, R.,** Transformation of mammalian cells with genes from procaryotes and eucaryotes, *Cell,* 16, 777, 1979.

21. **Maniatis, T., Fritsch, E. F., and Sambrook, J.,** *Molecular Cloning: A Laboratory Manual,* Cold Spring Harbor Press, Cold Spring Harbor, NY, 1982.

22. **Lastowski-Perry, D., Otto, E., and Maroni, G.,** Nucleotide sequence and expression of a *Drosophila* metallothionein, *J. Biol. Chem.,* 260, 1527, 1985.

23. **Griswold, D. E., Hillegas, L., Antell, L., Shatzman, A., and Hanna, N.,** Quantitative Western blot assay for measurement of the murine acute phase reactant, serum amyloid P component, *J. Immunol. Methods,* 91, 163, 1986.

24. **Hamer, D. H.,** Metallothionein, *Annu. Rev. Biochem.,* 55, 913, 1986.

25. **Pavlakis, G. N. and Hamer, D. H.,** Regulation of a metallothionein-growth hormone hybrid gene in bovine papilloma virus, *Proc. Natl. Acad. Sci. U.S.A.,* 80, 397, 1983.

26. **Barbacid, M.,** *ras* genes, *Annu. Rev. Biochem.,* 56, 779, 1987.

27. **Varmus, H. E.,** The molecular genetics of cellular oncogenes, *Annu. Rev. Genet.,* 18, 553, 1984.

28. **Pulciani, S., Santos, E., Long, L. K., Sorrentino, V., and Barbacid, M.,** *ras* gene amplification and malignant transformation, *Mol. Cell. Biol.,* 5, 2836, 1985.

29. **Chiang, L.-C., Silnutzer, J., Pipas, J. M., and Barnes, D. W.,** Selection of transformed cells in serum-free media, *In Vitro Cell. Dev. Biol.,* 21, 707, 1985.

30. **Yokoyama, S. and Sweet, R.,** unpublished data, 1985.

31. **Capon, D. J., Chen, E. Y., Levinson, A. D., Seeburg, P. H., and Goeddel, D. V.,** Complete nucleotide sequences of the T24 human bladder carcinoma oncogene and its normal homologue, *Nature (London),* 302, 33, 1983.

32. **Hagag, N., Halegoua, S., and Viola, M.,** Inhibition of growth factor-induced differentiation of PC12 cells by microinjection of antibody to *ras* p21, *Nature (London),* 319, 680, 1986.

33. **Kataoka, T., Powers, S., Cameron, S., Fasano, O., Goldfarb, M., Broach, J., and Wigler, M.,** Functional homology of mammalian and yeast *RAS* genes, *Cell*, 40, 19, 1985.
34. **Bishop, J. G. and Corces, V.,** Developmental consequences of overexpression of *RAS* and activated *RAS* in *Drosophila* germline transformants, presented at 3rd Annu. Meet. Oncol., Frederick, MD, July 7 to 11, 1987, 79.
35. **van der Straten, A., Johansen, H., Rosenberg, M., and Sweet, R.,** unpublished data, 1987.
36. **Shields, G. and Sang, J. H.,** Improved medium for culture of *Drosophila* embryonic cells, *Drosoph. Inf. Serv.*, 52, 161, 1977.
37. **Johansen, H., van der Straten, A., and Rosenberg, M.,** unpublished data, 1987.
38. **Reff, M.,** unpublished data, 1987.
39. **McGarry, T. J. and Lindquist, S.,** The preferential translation of *Drosophila* hsp70 mRNA requires sequences in the untranslated leader, *Cell*, 42, 903, 1985.

Chapter 21

EXPRESSION OF THE *AUTOGRAPHA CALIFORNICA* NUCLEAR POLYHEDROSIS VIRUS GENOME IN INSECT CELLS

Cornelia Oellig, Brigitte Happ, Thomas Müller, and Walter Doerfler

TABLE OF CONTENTS

I. INTRODUCTION

The insect baculovirus *Autographa californica* nuclear polyhedrosis virus (AcNPV) has become a well-studied model for investigations on the molecular biology of insect cells (for review, Doerfler and Böhm[1]). The viral genome is a supercoiled, circular, double-stranded DNA molecule of 126 to 129 kilobase pairs (kbp). The mode of transcription of the viral genome in insect cells, e.g., in *Spodoptera frugiperda* cells, a frequently used lepidopteran host cell line, is characterized by a number of peculiarities which are not observed for many other viral DNA genomes. First, the AcNPV-specific transcripts in insect cells do not seem to be extensively spliced.[2,3] It has not yet been clearly shown that insect cellular RNAs are extensively spliced in *S. frugiperda* cells. Second, the existence of nested sets of multiple overlapping RNA molecules has first been described in the *Eco*RI-J and -N segments of the AcNPV genome.[4] It has subsequently been demonstrated that this unique pattern of AcNPV transcripts is common to many, perhaps most, parts of the viral genome.[5-7] The biological significance of this interesting mode of viral transcription has not been clarified. Third, there is evidence that the different viral transcripts might be dependent on a complex temporal regulation of the expression program.[5,6,8,9]

In the present communication, the functional organization of the AcNPV genome between map coordinates 81.2 and 85.0 has been determined.

II. MATERIALS AND METHODS

The techniques applied in this study have been described previously.[3,4,6]

III. RESULTS

A. Restriction Map and Nucleotide Sequence Analyses of Parts of the *Eco*RI-N and -J Fragments of AcNPV DNA

The AcNPV isolate used in our laboratory was derived from a single plaque isolate[10] that was selected from a virus stock. The DNA from this AcNPV isolate differed slightly in its restriction map from that used in many other laboratories. The *Eco*RI-N and -J fragments in our DNA preparations correspond to map positions 79.3 to 86.4, or to the *Eco*RI-E and -H fragments in the conventional AcNPV map,[11] respectively.

The scheme in Figure 1a presents the *Eco*RI restriction map of AcNPV DNA;[3,12] Figure 1b, a detailed restriction map of the *Eco*RI-N and -J regions of the AcNPV genome. This map was obtained by standard restriction enzyme cleavage procedures. The map was determined in detail for the 81.2 to 85.0 map unit fragment across the *Eco*RI-N-J border, since these were the DNA segments from which the nested set of overlapping RNA molecules was transcribed.[4] The nucleotide sequence for the segment of AcNPV DNA shown in Figure 1b was determined by using M13 subclones or synthetic oligonucleotide primers in the standard dideoxy-chain termination procedure.[13] Individual subclones used in the sequencing experiments were designated by straight arrows (Figure 1b). Oligonucleotide primer-initiated sequences were designated by wiggly line arrows. The sequence of 4889 nucleotides, and the locations of the initiation and termination sites of the major open reading frames, were reproduced in Figure 2. The printout format of this sequence was based on the program SEQED.

B. Results of Computer Analyses of the Nucleotide Sequence Comprising Map Units 81.2 to 85.0 of the AcNPV Genome

The scheme in Figure 3 reproduced all open reading frames (ORFs) on the 5′ → 3′ strand in the nucleotide sequence that the computer program FRAMES had detected. At least in

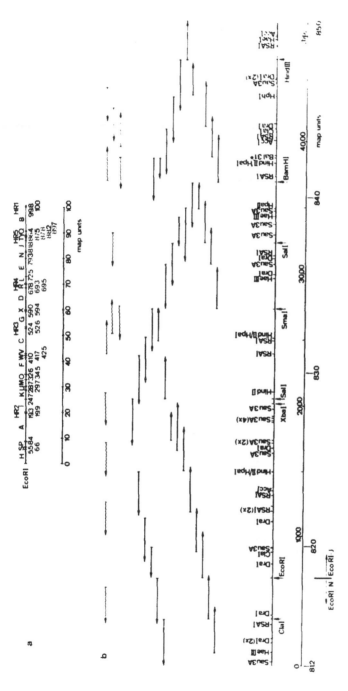

FIGURE 1.   (a) *EcoRI* restriction map of AcNPV DNA. The numbers refer to map units on a scale from 1 to 100. Conventionally, the AcNPV map, though circular in reality, is linearized;[11] (b) Detailed restriction map of the **81**.2- to 85.0-map unit segment of the viral genome. Only those restriction sites were indicated which were relevant for sequencing. Nucleotide numbers refer to the determined sequence (Figure 2). As will be apparent from this scheme, the nucleotide sequence of each subsegment was determined at least twice. Segments designated by straight line arrows were sequenced as M13 subclones using a commercial M13 primer. Segments marked by wiggly line arrows were sequenced by using synthetic oligonucleotide primers and the *EcoRI*-J fragment clone. (Reproduced from Oellig, C., Happ, B., Müller, T., and Doerfler, W., *J. Virol.*, 61, 3048, 1987. With permission.)

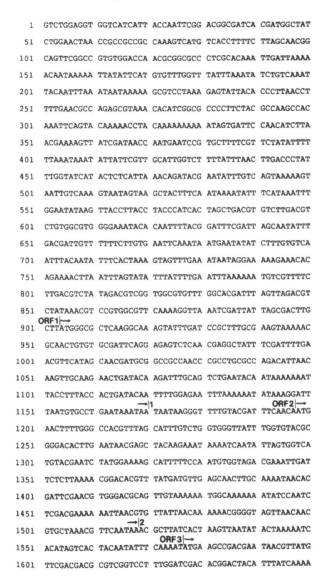

FIGURE 2.   Nucleotide sequence of the entire 81.2- to 85.0-map unit fragment of AcNPV DNA. The sequence of the 5′ → 3′ DNA strand (left → right) is presented. Initiation (| →) and termination (→ |) codons of presumptive ORFs were indicated. The two partly overlapping polyadenylation signals were underlined; the termination points were marked (*). (Reproduced from Oellig, C., Happ, B., Müller, T., and Doerfler, W., *J. Virol.*, 61, 3048, 1987. With permission.)

this section, the AcNPV genome did not contain major ORFs on the leftward-transcribed strand, with the possible exception of two relatively short ORFs. ORFs were found in all three reading frames on the rightward-transcribed strand, and there was no overlap between them (Figures 3 and 4). The common 3′ ends of these RNAs were determined by S1 protection analysis to nucleotide 4498 or to a position 8 bp downstream (see below). The individual 5′ ends were also assigned (cf. Figure 5). RNA molecules of lower abundance were designated by weaker lines. Map locations of the ORFs were compared with the exact positions of the nine major RNA size classes (cf. Figures 3 and 5). The positions of ORFs and RNAs relative to each other were consistent with the possibility that the selective translation of

```
1651  TTTAAAAATG  CCTTTGCAGG  CGTTTCAACA  ACTTTTGTTC  ACCATTCCAT

1701  CTAAACATAG  AAAAATGATC  AACGATGCGG  GCGGATCGTG  TCATAACACG

1751  GTCAAATACA  TGGTGGACAT  TTACGGAGCG  GCCGTTCTGG  TTTTGCGAAC

1801  GCCTTGCTCG  TTCGCCGACC  AGTTGTTGAG  CACATTTATT  GCAAACAATT

1851  ATTTGTGCTA  CTTTTACCGT  CGTCGCCGAT  CACGATCACG  CTCACGATCA

1901  CGCTCGCGAT  CACGTTCTCC  TCATTGCAGA  CCTCGTTCGC  GCTCTCCTCA

1951  TTGCAGACCT  CGTTCGCGAT  CTCGGTCCCG  GTCTAGATCG  CGGTCACGTT

2001  CATCGTCTCC  CAGGCGAGGG  CGTCGACAAA  TATTCGACGC  GCTGGAAAAG

2051  ATTCGTCATC  AAAACGACAT  GTTGATGAGC  AACGTCAACC  AAATAAATCT

2101  CAACCAAACT  AATCAATTTT  TAGAATTGTC  CAACATGATG  ACGGGCGTGC

2151  GCAATCAAAA  CGTGCAGCTC  CTCGCGGCGT  TGGAAACCGC  TAAAGATGTT

2201  ATTTTGACCA  GATTAAACAC  ATTGCTTGCC  GAGATTACAG  ACTCGTTACC

2251  CGACTTGACG  TCCATGTTAG  ATAAATTAGC  TGAACAATTG  TTGGACGCCA

2301  TCAACACGGT  GCAGCAAACG  CTGCGCAACG  AGTTGAACAA  CACCAACTCT

2351  ATTTTGACCA  ATTTAGCGTC  AAGCGTCACA  AACATCAACG  GTACGCTCAA

2401  CAATTTGCTA  GCCGCTATCG  AAAACTTAGT  AGGCGGCGGC  GGCGGTGGCA

2451  ATTTTAACGA  AGCCGACAGA  CAAAAACTGG  ACCTCGTGTA  CACTTTGGTT

                                                    ORF4|→
2501  AACGAAATCA  AAAATATACT  CACGGGAACG  CTGACAAAAA  AATAAGCATG
                                                      →|3

2551  TCCGACAAAA  CACCAACAAA  AAAGGGTGGC  AGCCATGCCA  TGACGTTGCG

2601  AGAGCGCGGC  GTAACAAAAC  CCCCAAAAAA  GTCTGAAAAG  TTGCAGCAAT

2651  ACAAGAAAGC  CATCGCTGCC  GAGCAAACGC  TGCGCACCAC  AGCAGATGTT

2701  TCTTCTTTGC  AGAACCCCGG  GGAGAGTGCC  GTTTTTCAAG  AGTTGGAAAG

2751  ATTAGAGAAT  GCAGTTGTAG  TATTAGAAAA  TGAACAAAAA  CGATTGTATC

2801  CCATATTAGA  TACGCCTCTT  GATAATTTTA  TTGTCGCATT  CGTGAATCCG

2851  ACGTATCCCA  TGGCCTATTT  TGTCAATACC  GATTACAAAT  TAAAACTAGA

2901  ATGTGCCAGA  ATCAGAAGCG  ATTTACTTTA  CAAAAACAAA  AACGAAGTCG

2951  CTATCAACAG  GCCTAAGATA  TCGTCTTTTA  AATTGCAATT  GAACAACGTA

3001  ATTTTAGACA  CTATAGAAAC  TATTGAATAC  GATTTACAAA  ATAAAGTTCT

3051  CACAATTACT  GCACCTGTTC  AAGATCAAGA  ACTAAGAAAA  TCCATTATTT

3101  ATTTTAATAT  TTTAAATAGT  GACAGTTGGG  AAGTACCAAA  GTATATGAAA

                  →|4                            ORF5|→
3151  AAATTGTTTG  ATGAAATGCA  ATTGGAACCT  CCCGTCATTT  TACCATTAGG

3201  TCTTTAGATT  TGGTAAGGCT  AGCACGTCGA  CATCATGTTT  GCGTCGTTGA

3251  CCTCAGAGCA  AAAGCTGTTA  TTAAAAAAAT  ATAAATTTAA  CAATTATGTG
```

FIGURE 2 continued

individual size classes of RNAs would provide a different and unique set of AcNPV-specific polypeptides in infected cells. Such a mechanism would introduce variability into the expression pattern, although it remained unknown which factors influenced the selection for translation of individual RNA size classes.

The data compiled in Figure 3 also demonstrate that the nucleotide sequences preceding many of the ORFs carry TATA-like signals and presumptive CAAT boxes: in particular, the TATA and CAAT boxes in front of the sites of initiation of the 3341, and 1948 nucleotide RNAs were located at the expected distances from the actual cap sites (Figure 3, TATA and CAAT signals designated by arrows). The nucleotide sequences at the 5' ends around the ATG initiation triplets of some of the ORFs conform to the Kozak rules.[14] Enhancer-consensus sequences[15] have not been detected. Perhaps the enhancing signals would be located more proximally or more distally, possibly in one of the neighboring HR regions[16] (cf. Figure 1A), although these HR regions are very distant. The nucleotide sequence reveals

```
3301  AAAACGATCG AGTTGAGTCA AGCGCAGTTG GCTCATTGGC GTTCAAACAA
3351  AGATATTCAG CCAAAACCTT TGGATCGTGC AGAAATTTTA CGTGTCGAAA
3401  AGGCCACCAG GGGACAAAGC AAAAATGAGC TGTGGACGCT ATTGCGTTTG
3451  GATCGCAACA CAGCGTCTGC ATCGTCCAAC TCGTCCGGCA ACATGTTACA
3501  ACGACCAGCG CTTTTGTTTG GAAACGCGCA AGAAAGTCAC GTCAAAGAAA
3551  CCAACGGCAT CATGTTAGAC CACATGCGCG AAATCATAGA AAGTAAAATT
3601  ATGAGCGCGG TCGTTGAAAC GGTTTTGGAT TGCGGCATGT TCTTTAGCCC
3651  CTTGGGTTTG CACGCCGCTT CGCCCGATGC GTATTTTTCT CTCGCCGACG
3701  GAACGTGGAT CCCAGTGGAA ATAAAATGTC CGTACAATTA CCGAGACACG
3751  ACCGTGGAGC AGATGCGTGT CGAGTTGGGG AACGGCAATC GCAAGTATCG
3801  CGTGAAACAC ACCGCGCTGT TGGTTAACAA GAAAGGCACG CCCCAGTTCG
3851  AAATGGTCAA AACGGATGCG CATTACAAGC AAATGCAACG GCAGATGTAT
3901  GTGATGAACG CGCCTATGGG CTTTTTACGTG GTCAAATTCA AACAAAATTT
3951  GGTGGTGGTT TCTGTGCCGC GCGACGAAAC GTTCTGCAAC AAAGAACTGT
4001  CTACGGAAAA CAACGCGTAC GTGGCGTTTG CCGTGGAAAA CTCCAACTGC
4051  GCGCGCTACC AATGCGCCGA CAAGCGACGG CTTTCATTCA AAACGCACAG
4101  CTGCAATCAC AACTATAGTG GTCAAGAAAT CGATGCTATG GTCGATCGCG
4151  GAATATATTT AGATTATGGA CATTTAAAAT GTGCGTACTG TGATTTTAGC
4201  TCAGACAGTC GGGAAACGTG CGATTCTGTT TTAAAACGCG AGCACACCAA
4251  CTGCAAAAGT TTTAACTTGA AACATAAAAA CTTTGACAAT CCTACATACT
4301  TTGATTATGT TAAAAGATTG CAAAGTTTGC TAAAGAGTCA CCACTTTAGA
4351  AACGACGCTA AAACACTTGC CTATTTTGGT TACTATTTAA CTCATACAGG
4401  AACCCTGAAG ACCTTTTGCT GCGGATCGCA AAACTCGTCG CCCACCAAAC
4451  ACGATCATTT AAACGACTGT GTATATTATT TGGAAATAAA ATAAACCTTT
4501  ATATTATATA TAATTCTTTT ATTTATACAT TTGTTTATAC AATTTTATTT
4551  ACGACAAATA TTGACTCGTT GTTCAGAAAG TTTAATAAGC TTGTCAATTT
4601  CTTCGGCTTG CAAAGGGCTG CCAACGCGTT CGTTTTGAAT GCGCGTAATC
4651  CGGTTTACGG TATTGTTGGC GCGAACAATA AACTCCTCAA CTGGCAAATT
4701  AACATTTTGT TTGCGTACTC ATTGTGCACT GCGGCCAGGT TTTGTAGAAT
4751  GTTTTCGGGA AAAATGGCAA TTCTATTAAA TTTGACATGT TTTTGATTGT
4801  ATACATAGTT TTGATATTCT TCCAGCGTAG GATATTTGTT TAAACTCTTG
4851  ACGCATTCAA TGTACAATTT GTGCAGTGAC AAAATTCTG
```

Above position 4451, the following markers appear:
```
                                        *  →|5  *
                              TGGAAATAAA ATAAACCTTT
```

FIGURE 2 continued

several potential polyadenylation signals. Three of these sites, which lie close to the common 3′ end of the RNAs, may be of functional significance. These latter polyadenylation signals are followed in their 3′ vicinity by TGTT-containing sequences which have been previously identified downstream from polyadenylation sites.[17] Moreover, the scheme in Figure 3 lists the number of nucleotides in each sequence comprising an ORF and the molecular weights (in kilodaltons) of the presumptive polypeptides encoded within these ORFs. It is worth recalling that in an earlier study, polypeptides of 34K (major component), 48K, and 28K (minor components) were observed after *in vitro* translation of RNA molecules from AcNPV-infected cells which had been hybrid selected on the cloned *Eco*RI-J fragment of the AcNPV genome.[3] Similar *in vitro* translation results were obtained with RNAs hybrid-selected on subfragments of the *Eco*RI-J fragment (see below).

C. Amino Acid Sequences of the Presumptive Proteins in Open Reading Frames 1 to 5

In Figure 4, the amino acid sequences of the five ORFs in the 81.2 to 85.0 region of the

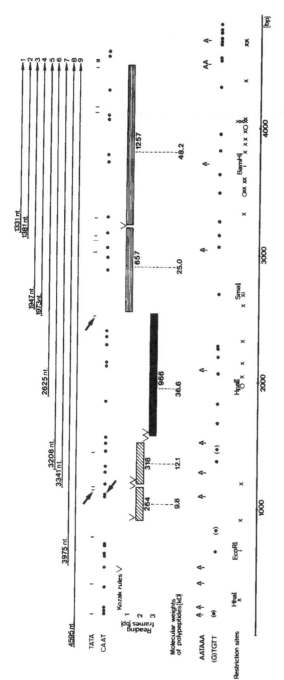

FIGURE 3. Genetic signals in the 81.2- to 85.0-map unit segment of the AcNPV genome. Most of the signals were explained in the scheme itself or in the text. The nine major size classes of RNAs (1 to 9) were designated by arrows, and their lengths were indicated in nucleotides. The ORFs in frames 1 to 3 were designated ▤ in reading frame 1, ▨ in reading frame 2, ■ in reading frame 3. Some of the important restriction sites were also included (cf. Figure 1b). In Figures 3 and 5 (see below), RNAs of lesser abundance were designated by weaker lines. (Reproduced from Oellig, C., Happ, B., Müller, T., and Doerfler, W., *J. Virol.*, 61, 3048, 1987. With permission.)

ORF1: MetGlyAlaGlnGlyLysValPheAspProLeuCysGluValLysThrGlnLeuCysAla
IleGlnGluSerLeuAsnGluAlaIleSerIleLeuAsnValHisSerAsnAspAlaAla
AlaAsnProProAlaProAspIleAsnLysLeuGlnGluLeuIleGlnAspLeuGlnSer
GluTyrAsnLysLysIleThrPheThrThrAspThrIleLeuGluAsnLeuLysAsnIle
LysAspLeuMetCysLeuAsnLys

ORF2: MetAsnPheTrpAlaThrPheSerIleCysLeuValGlyTyrLeuValTyrAlaGlyHis
LeuAsnAsnGluLeuGlnIleLysSerIleLeuValValMetTyrGluSerMetGlu
LysHisPheSerAsnValValAspGluIleAspSerLeuLysThrAspThrPheMetMet
LeuSerAsnLeuGlnAsnAsnThrIleArgThrTrpAspAlaValValLysAsnGlyLys
LysIleSerAsnLeuAspGluLysIleAsnValLeuLeuThrLysAsnGlyValValAsn
AsnValLeuAsnValGln

ORF3: MetLysProThrAsnAsnValMetPheAspAspAlaSerValLeuTrpIleAspThrAsp
TyrIleTyrGlnAsnLeuLysMetProLeuGlnAlaPheGlnGlnLeuLeuPheThrIle
ProSerLysHisArgLysMetIleAsnAspAlaGlyGlySerCysHisAsnThrValLys
TyrMetValAspIleTyrGlyAlaAlaValLeuValLeuArgThrProCysSerPheAla
AspGlnLeuLeuSerThrPheIleAlaAsnAsnTyrLeuCysTyrPheTyrArgArgArg
ArgSerArgSerArgSerArgSerArgSerArgSerArgSerProHisCysArgProArg
SerArgSerProHisCysArgProArgSerArgSerArgSerArgSerArgSerArgSer
ArgSerSerSerProArgArgGlyArgArgGlnIlePheAspAlaLeuGluLysIleArg
HisGlnAsnAspMetLeuMetSerAsnValAsnGlnIleAsnLeuAsnGlnThrAsnGln
PheLeuGluLeuSerAsnMetMetThrGlyValArgAsnGlnAsnValGlnLeuLeuAla
AlaLeuGluThrAlaLysAspValIleLeuThrArgLeuAsnThrLeuLeuAlaGluIle
ThrAspSerLeuProAspLeuThrSerMetLeuAspLysLeuAlaGluGlnLeuLeuAsp
AlaIleAsnThrValGlnGlnThrLeuArgAsnGluLeuAsnAsnThrAsnSerIleLeu
ThrAsnLeuAlaSerSerValThrAsnIleAsnGlyThrLeuAsnAsnLeuAlaAlaAla
IleGluAsnLeuValGlyGlyGlyGlyGlyGlyAsnPheAsnGluAlaAspArgGlnLys
LeuAspLeuValTyrThrLeuValAsnGluIleLysAsnIleLeuThrGlyThrLeuThr
LysLys

ORF4: MetSerAspLysThrProThrLysLysGlyGlySerHisAlaMetThrLeuArgGluArg
GlyValThrLysProProLysLysSerGluLysLeuGlnGlnTyrLysLysAlaIleAla
AlaGluGlnThrLeuArgThrThrAlaAspValSerSerLeuGlnAsnProGlyGluSer
AlaValPheGlnGluLeuGluArgLeuGlnAsnAlaValValValLeuGluAsnGluGln
LysArgLeuTyrProIleLeuAspThrProLeuAspAsnPheIleValAlaPheValAsn
ProThrTyrProMetAlaTyrPheValAsnThrAspTyrLysLeuLysLeuGluCysAla
ArgIleArgSerAspLeuLeuTyrLysAsnLysAsnGluValAlaIleAsnArgProLys
IleSerSerPheLysLeuGlnLeuAsnAsnValIleLeuAspThrIleGluThrIleGlu
TyrAspLeuGlnAsnLysValLeuThrIleThrAlaProValGlnAspGlnGluLeuArg
LysSerIleIleTyrPheAsnIleLeuAsnSerAspSerTrpGluValProLysTyrMet
LysLysLeuPheAspGluMetGlnLeuGluProProValIleLeuProLeuGlyLeu

ORF5: MetPheAlaSerLeuThrSerGluGlnLysLeuLeuLeuLysLysTyrLysPheAsnAsn
TyrValLysThrIleGluLeuSerGlnAlaGlnLeuAlaHisTrpArgSerAsnLysAsp
IleGlnProLysProLeuAspArgAlaGluIleLeuArgValGluLysAlaThrArgGly
GlnSerLysAsnGluLeuTrpThrLeuLeuArgLeuAspArgAsnThrAlaSerAlaSer
SerAsnSerSerGlyAsnMetLeuGlnArgProAlaLeuLeuPheGlyAsnAlaGlnGlu
SerHisValLysGluThrAsnGlyIleMetLeuAspHisMetArgGluIleIleGluSer
LysIleMetSerAlaValValGluThrValLeuAspCysGlyMetPhePheSerProLeu
GlyLeuHisAlaAlaSerProAspAlaTyrPheSerLeuAlaAspGlyThrTrpIlePro
ValGluIleLysCysProTyrAsnTyrArgAspThrThrValGluGlnMetArgValGlu
LeuGlyAsnGlyAsnArgLysTyrArgValLysHisThrAlaLeuLeuValAlaAsnLysLys
GlyThrProGlnPheGluMetValLysThrAspAlaHisTyrLysGlnMetGlnArgGln
MetTyrValMetAsnAlaProMetGlyPheTyrValValLysPheLysGlnAsnLeuVal
ValValSerValProArgAspGluThrPheCysAsnLysGluLeuSerThrGluAsnAsn
AlaTyrValAlaPheAlaValGluAsnSerAsnCysAlaArgTyrGlnCysAlaAspLys
ArgArgLeuSerPheLysThrHisSerCysAsnHisAsnTyrSerGlyGlnGluIleAsp
AlaMetValAspArgGlyIleTyrLeuAspTyrGlyHisLeuLysCysAlaTyrCysAsp
PheSerSerAspSerArgGluThrCysAspSerValLeuLysArgGluHisThrAsnCys
LysSerPheAsnLeuLysHisLysAsnPheAspAsnProThrTyrPheAspTyrValLys
ArgLeuGlnSerLeuLeuLysSerHisHisPheArgAsnAspAlaLysThrLeuAlaTyr
PheGlyTyrTyrLeuThrHisThrGlyThrLeuLysThrPheCysCysGlySerGlnAsn
SerSerProThrLysHisAspHisLeuAsnAspCysValTyrTyrLeuGluIleLys

FIGURE 4.   Amino acid sequences of the presumptive polypeptides encoded by ORFs 1 to 5. These sequences were derived from the nucleotide sequence presented in Figure 2 by the computer programs MAP and PUBLISH. The numbering of the ORFs corresponded to that in Figure 3. The striking repeat of ARG-SER sequences in ORF3 was underlined. (Reproduced from Oellig, C., Happ, B., Müller, T., and Doerfler, W., *J. Virol.*, 61, 3048, 1987. With permission.)

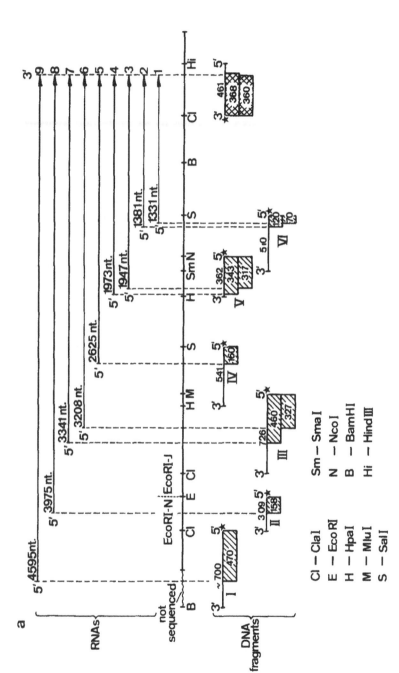

FIGURE 5. Mapping of the 5' termini of nine different RNA size classes in the region of the *EcoRI-N/J* fragments of AcNPV DNA. Experimental conditions were described in the text. Figure shows schematic description of the mapping experiments. In the center of the scheme, the relevant restriction sites (as explained in the graph) were indicated. The DNA fragments indicated with sizes in base pairs (bp) (DNA and bracketed fragments I to VI) were all [32P]-labeled at the 5' termini (*). The subfragments designated by hatched boxes and identified by nucleotide lengths were protected by RNA annealing in S1 nuclease protection experiments. The subfragments, which were preserved in the S1 cleavage experiment and which were used to determine the 3' termini, were also included as cross-hatched boxes. In the upper part of the scheme (RNAs and bracket), the nine major RNA size classes were schematically shown with their lengths and exact map locations. (Reproduced from Oellig, C., Happ, B., Müller, T., and Doerfler, W., *J. Virol.*, 61, 3048, 1987. With permission.)

AcNPV genome were reproduced. These ORFs comprised potential polypeptides of 9.8, 12.1, 25.0, 36.6, and 48.2 kDa. A potentially very interesting polypeptide was that in ORF3. Its sequence exhibited a 16-fold repeat of the sequence ARG-SER in a polypeptide in which the basic amino acids predominated. Due to differences in codon usage, these repeats in the amino acid sequence were not reflected in the nucleotide sequence. A symmetry was recognized in the arrangement . . . ARG-ARG-ARG-SER- . . . ARG-SER-SER-SER . . . at the flanks of this repeat. The ARG-SER repeat was subdivided into two segments which were separated by two repeats of the sequence PRO-HIS-CYS-ARG-PRO. The actual occurrence of any of the theoretically possible polypeptides in AcNPV-infected *S. frugiperda* cells remained as yet unproven.

D. Determination of the 3′ Termini of the RNA Molecules by S1 Protection Analysis

The 461 nucleotide *Cla*I-*Hin*dIII subfragment, as indicated in the scheme in Figure 5, was labeled at the 3′ termini by filling in α-[^{32}P] deoxyribonucleotides using the Klenow fragment of the DNA polymerase. This fragment was used for S1 protection analyses of the 3′ termini of the different RNA size classes. The DNA fragment protected by hybridization to RNA from AcNPV-infected *S. frugiperda* cells against S1 nuclease cleavage had a length of 360 nucleotides (Figure 5). Based on the sequence data presented in Figure 2, the 3′ termini of the RNA molecules could, thus, be placed at nucleotide 4490. A second major fragment of 368 nucleotides in length was also observed demonstrating that a second 3′ end of, at least, some of the RNA molecules was located at nucleotide 4498. There possibly existed a third minor band of an intermediate length. It was concluded that within a sequence of eight nucleotides, all nine RNA size classes terminated at the same site.

E. Mapping of the Individual 5′ Termini of RNA Molecules by S1 Protection Analyses

The scheme in Figure 5 presented the map locations and sites of labeling (designated by *) of six restriction subfragments (I to VI) in the *Eco*RI-N/J regions of AcNPV DNA. These subfragments were used in assessing the individual 5′ start sites of the nine major RNA size classes. Map locations of these RNAs were also indicated. The exact protocols of preparing and labeling the designated fragments were described earlier.[6] The sizes of the DNA fragments protected in S1 nuclease experiments were 470 nucleotides (nt) for fragment I, 158 nt for fragment II, 460 and 327 nt for fragment III, 160 nt for fragment IV, 343 and 317 nt for fragment V, 120 and 70 nt for fragment VI (data not shown; Oellig et al.[6]). We observed a multitude of protected minor subfragments as one progressed from fragments II to VI in the protection experiments. These results suggested the existence of multiple, less well-defined start sites. Could this increase in start sites be explained by the occurrence of numerous additional, minor start sites of transcription in the region of the *Eco*RI-J fragment? The 368- and 360-nt fragments protected by all RNAs at the 3′ termini in the *Cla*I-*Hin*dIII fragment were also shown in Figure 5.

The fragment lengths presented placed the sites of transcription initiation for RNA size classes 1 through 9 to nucleotides 105 (this part of the DNA segment was not sequenced), 515, 1149, 1282, 1865, 2517, 2543, 3109, and 3159, respectively. These numbers referred to the nucleotide sequence reproduced in Figure 2. The locations of the 5′ termini of the nine RNA size classes shown schematically in Figures 3 and 5 conform to these determinations. The lengths of the different size classes are 4595, 3975, 3341, 3208, 2625, 1973, 1947, 1381, and 1331 nt.

F. *In vitro* Translation of Viral RNAs Hybrid-Selected on Subfragments of the *Eco*RI-J Fragment of AcNPV DNA

Previous *in vitro* translation analyses of viral RNAs were performed with RNAs hybrid-selected on the entire *Eco*RI-J fragment.[3] It was conceivable that some of the RNAs selected in this way might have originated in neighboring fragments. Moreover, by using subfragments of the *Eco*RI-J fragment for hybrid-selection experiments, the analysis could be improved in sensitivity. Cytoplasmic RNA was isolated 24 h after infection of *S. frugiperda* cells with AcNPV. AcNPV-specific RNAs were hybrid-selected on the *Eco*RI-J fragment, or on the *Eco*RI-*Sma*I, or on the *Sma*I-*Bam*HI subfragment. The locations of these subfragments were apparent from the maps in Figures 3 and 5. The selected RNAs were *in vitro* translated in a reticulocyte cell-free lysate by standard procedures. The *in vitro* synthesized polypeptides were labeled with a mixture of [³H]-labeled amino acids and analyzed by electrophoresis on SDS polyacrylamide gels. When the entire *Eco*RI-J fragment was used for RNA selection, polypeptides of 48 to 50, 34 to 36, and 28 kDa were observed. With the *Eco*RI-*Sma*I fragment, RNAs encoding a 34- to 36- and a 28-kDa polypeptide were selected, and the 48- to 50-kDa protein was hardly detectable with this RNA population. Upon RNA selection on the *Sma*I-*Bam*HI fragment, the three translation products listed above were again obtained.

These results confirmed earlier findings and indicated which of the available ORFs were actually utilized. The sizes of the polypeptides synthesized *in vitro* under the control of hybrid-selected RNAs were fully commensurate with the lengths of some of the ORFs read off the nucleotide sequence, i.e., the presumptive proteins with molecular weights of 48.2, 36.6, and 25.0 kDa (Figures 3 and 4). Polypeptides of 9.8 or 12.1 kDa were not seen in *in vitro* translation experiments.

IV. DISCUSSION

A. Nucleotide Sequence Confirms Previously Assigned Transcriptional Arrangement in the *Eco*RI-J and -N Fragments of AcNPV DNA

By nucleotide sequence determination and S1 protection analyses, the 3' and 5' termini of the nine major size classes of AcNPV-specific RNAs have been precisely mapped relative to the *Eco*RI-J and -N fragments (Figures 2, 3, and 5). The results have refined the previously assigned map positions of these RNAs.[4] The 3' ends of the nested sets of RNAs correspond to nucleotide numbers 4490 and 4498 of the sequence determined. There are two poly-adenylation sites close to and one following the common 3' terminus of the nine major RNA size classes. The 5' ends of the individual RNA size classes are placed at the following nucleotides: 105, 515, 1149, 1282, 1865, 2517, 2543, 3109, and 3159 in the determined sequence (Figures 2 and 5). In this way the map locations of these RNAs have been unequivocally determined.

As mentioned above, there is evidence that several additional RNA size classes, apart from the nine major classes mapped precisely, are transcribed from the *Eco*RI-J fragment of AcNPV DNA in AcNPV-infected *S. frugiperda* cells. This finding might indicate that the mechanism of transcription initiation was not very precise but, rather, allowed initiations at quite a number of sites which were remarkably rich in A-T nucleotide pairs. The nine determined sites (Figure 5) were the predominant ones.

B. Sequential Array of Polypeptides in the *Eco*RI-J Fragment

On the rightward-transcribed strand of the linearized AcNPV genome,[11] five ORFs could be recognized which were located in different reading frames (Figures 2 to 4). The corresponding polypeptides ranged in sizes from 9.8 to 48.2 kDa and were serially arranged in this part of the viral genome without any overlap (Figure 3). It was not yet proven that all these polypeptides were synthesized in AcNPV-infected *S. frugiperda* cells. The arrangement of ORFs was suggestive in that it would allow for the step-wise sequential expression of certain combinations of viral polypeptides, depending on which size classes of RNAs from this region were predominantly synthesized and chosen for translation. In this way, the viral expression schedule might be adapted to specific needs in the viral replication cycle. It was proposed earlier[4] that the flexibility introduced by this mode of viral gene expression might be a substitute for the versatility afforded by the splicing mechanism that other viral systems, but less so AcNPV, have perfected. There may be additional, more complicated explanations for the occurrence of multiple RNA molecules in many sequences of the AcNPV genome. Moreover, AcNPV infection might somehow interfere with cellular RNA splicing and shut off cellular gene expression, since AcNPV gene expression was apparently rendered independent of the splicing mechanism.

Among the presumptive ORFs, a potentially interesting polypeptide is the 36.6-kDa protein which contains a 16-fold repeat of the ARG-SER dipeptide sequence. It will be challenging to investigate whether this protein is actually synthesized in AcNPV-infected insect cells. Since a 34- to 36-kDa protein has also been found by *in vitro* translation of RNA selected on the *Eco*RI-J fragment or on some of its subfragments, it is likely that this polypeptide ranks among the viral gene products.

C. *In Vitro* Translation of Hybrid-Selected RNAs

The following polypeptides have been synthesized *in vitro* by translating AcNPV-specific RNAs selected on the *Eco*RI-*Sma*I and *Sma*I-*Bam*HI subfragments of the *Eco*RI-J fragment: a 48- to 50-, a 34- to 36-, and a 28-kDa polypeptide. It should be noted that the major ORFs as derived from the nucleotide sequence data agree quite well with the results of these and previous[3] *in vitro* experiments. *In vitro* translation experiments in reticulocyte lysates have often yielded reliable results and reflect, to some degree, the patterns of polypeptides synthesized in virus-infected cells. However, this method does not assure congruence of translation patterns *in vivo* and *in vitro*.

ACKNOWLEDGMENTS

We are indebted to Gerti Meyer zu Altenschildesche for media preparation, and to Petra Böhm for expert editorial work. This research was supported by the Deutsche Forschungsgemeinschaft through SFB74-C1, and by grant Do 165-2/1.

REFERENCES

1. **Doerfler, W. and Böhm, P., Eds.,** The molecular biology of baculoviruses, *Current Topics in Microbiology and Immunology,* Vol. 131, Springer-Verlag, Berlin, 1986.
2. **Howard, S. C., Ayres, M. D., and Possee, R. D.,** Mapping the 5' and 3' ends of *Autographa californica* nuclear polyhedrosis virus polyhedrin mRNA, *Virus Res.,* 5, 109, 1986.
3. **Lübbert, H. and Doerfler, W.,** Mapping of early and late transcripts encoded by the *Autographa californica* nuclear polyhedrosis virus genome: is viral RNA spliced?, *J. Virol.,* 50, 497, 1984.
4. **Lübbert, H. and Doerfler, W.,** Transcription of overlapping sets of RNAs from the genome of *Autographa californica* nuclear polyhedrosis virus: a novel method for mapping RNAs, *J. Virol.,* 52, 255, 1984.
5. **Friesen, P. D. and Miller, L. K.,** Temporal regulation of baculovirus RNA: Overlapping early and late transcripts, *J. Virol.,* 54, 392, 1985.
6. **Oellig, C., Happ, B., Müller, T., and Doerfler, W.,** Overlapping sets of viral RNAs reflect array of polypeptides in the *Eco*RI-J and -N fragments (map positions 81.2 to 85.0) of the *Autographa californica* nuclear polyhedrosis virus genome, *J. Virol.,* 61, 3048, 1987.
7. **Rankin, C., Ladin, B. F., and Weaver, R. F.,** Physical mapping of temporally regulated, overlapping transcripts in the region of the 10K protein gene in *Autographa californica* polyhedrosis virus, *J. Virol.,* 57, 18, 1986.
8. **Mainprize, T. H., Lee, K., and Miller, L. K.,** Variation in the temporal expression of overlapping baculovirus transcripts, *Virus Res.,* 6, 85, 1986.
9. **Rohel, D. Z. and Faulkner, P.,** Time course analysis and mapping of *Autographa californica* nuclear polyhedrosis virus transcripts, *J. Virol.,* 50, 739, 1984.

10. **Tjia, S. T., Carstens, E. B., and Doerfler, W.**, Infection of *Spodoptera frugiperda* cells with *Autographa californica* nuclear polyhedrosis virus. II. The viral DNA and the kinetics of its replication, *Virology*, 99, 399, 1979.
11. **Vlak, J. M. and Smith, G. E.**, Orientation of the genome of *Autographa californica* nuclear polyhedrosis virus: a proposal, *J. Virol.*, 41, 1118, 1982.
12. **Lübbert, H., Kruczek, I., Tjia, S. T., and Doerfler, W.**, The cloned *Eco*RI fragments of *Autographa californica* nuclear polyhedrosis virus DNA, *Gene*, 16, 343, 1981.
13. **Sanger, F., Nicklen, S., and Coulson, A. R.**, DNA sequencing with chain-terminating inhibitors, *Proc. Natl. Acad. Sci. U.S.A.*, 74, 5463, 1977.
14. **Kozak, M.**, Comparison of initiation of protein synthesis in procaryotes, eucaryotes and organelles, *Microbiol. Rev.*, 47, 1, 1983.
15. **Hearing, P. and Shenk, T.**, The adenovirus type 5 E1A transcriptional control region contains a duplicated enhancer element, *Cell*, 33, 695, 1983.
16. **Guarino, L. A., Gonzalez, M. A., and Summers, M. D.**, Complete sequence and enhancer function of the homologous DNA regions of *Autographa californica* nuclear polyhedrosis virus, J. Virol., 60, 224, 1986.
17. **McLauchlan, J., Gaffney, D., Whitton, H. L., and Clements, J. B.**, The consensus sequence YGTGTTYY located downstream from the AATAAA signal is required for efficient formation of mRNA 3' termini, *Nucleic Acids Res.*, 13, 1347, 1985.

Chapter 22

ENHANCERS OF EARLY GENE EXPRESSION

Linda A. Guarino

TABLE OF CONTENTS

I. INTRODUCTION

Expression of viral genes in AcNPV-infected cells is temporally regulated.[1,2] Four general classes of genes are recognized: the immediate early (α) genes, which are expressed in the absence of prior viral protein synthesis; the delayed early (β) genes, which require the synthesis of one or more α genes; the late (γ) genes which are concomitant with DNA synthesis; and the very late (δ) genes which are maximally synthesized during the occlusion phase. At least some of this temporal control is exerted at the level of transcription, as evidenced by the presence of different RNA transcripts at different times of infection.[3-5] Transcription of several β genes is transactivated by the α gene IE-1.[6,7] Activation of the late class of transcripts may be linked to a viral-induced RNA polymerase.[8]

The genome of the baculovirus *Autographa californica* nuclear polyhedrosis virus (AcNPV) is a circular double-stranded molecule of approximately 128 kilobase pairs (kbp). Although the genome consists primarily of unique sequences, five regions which share homologous sequences are interspersed along the length of the genome.[9] A similar pattern of interspersed homologous DNA is also found in another baculovirus, *Choristoneura fumiferana* nuclear polyhedrosis virus.[10] Here we report the nucleotide sequence of the AcNPV homologous regions and present evidence that they function as enhancers of delayed early gene transcription.

II. MATERIALS AND METHODS

A. Analysis of Enhancer Function

The conditions for cell culture, transfection, and CAT assays have been described.[6] To determine whether the homologous regions (*hr*s) function as enhancers, the following fragments were repaired with Klenow according to standard procedures[11] and cloned in both orientations into the *Hin*dIII (Klenow-repaired) site in the multiple cloning region upstream of 39CAT: for *hr*1, the *Cla*I fragment shown in Figure 1; for *hr*2, a 1.5-kb *Hin*dIII-*Sal*I fragment; for *hr*3, the *Mlu*I-*Ssp*I fragment shown in Figure 1, and for *hr*5, the *Mlu*I fragment shown in Figure 1. *Hr*4*left* was cloned in one orientation using an *Nsi*I-*Xba*I fragment into the *Pst*I and *Hin*dIII sites of 39CAT. The resulting plasmid contains *hr*4 in an orientation opposite to that of the standard AcNPV genetic map.

B. DNA Sequencing

The sequencing strategy for the five regions is indicated in Figure 1. The AcNPV restriction fragments *Pst*I-D (*hr*1), *Hin*dIII-L (*hr*2), *Hin*dIII-B (*hr*3), *Bgl*II-G (*hr*4), or *Hin*dIII-Q (*hr*5) were partially or completely digested with *Eco*RI, and ligated with *Eco*RI-digested M13mp9.[12] Single-stranded recombinant phage was purified and sequenced by the dideoxy-chain-termination procedure.[13] The restriction map data generated by initial sequencing was used to design additional cloning and sequencing to determine the order of the *Eco*RI minifragments. For *hr*2, *Bal*31 deletions of *Hin*dIII-L were constructed according to standard procedures[11] and sequenced to determine the order of the minifragments. The sequences were compiled and analyzed using the programs of Devereaux et al.[14]

III. RESULTS

A. Homologous Region DNAs Stimulate 39CAT Expression

Recently we reported that an immediate early gene (IE-1) of AcNPV *trans*-activates expression of chloramphenicol acetyltransferase under the control of the promoter for the delayed early 39K gene.[6] The genomic location of IE-1 was functionally mapped using a series of deletions of a plasmid containing the *Eco*RI-B fragment of AcNPV DNA. The

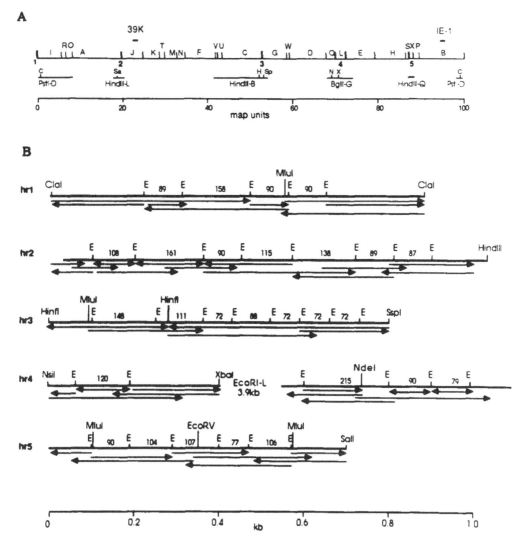

FIGURE 1. The arrangement of homologous sequences on the AcNPV genome and the sequencing strategy of the *hrs*. (A) The location of the *hrs* in the *Eco*RI restriction map is indicated by the heavy vertical lines numbered 1 to 5. The indicated clones which overlap the *hrs* were subcloned for sequencing. Some relevant restriction sites are indicated: C, *Cla*I; H, *Hin*dIII; Ss, *Ssp*I; N, *Nsi*I; X, *Xba*I; Sa, *Sal*I. By convention the genetic map of AcNPV is presented with the left end of *Eco*RI-I at position 0 (32); (B) the indicated restriction fragments were sequenced. In *hr2 Bal*31 deletions were constructed and sequenced. The number of bases between the *Eco*RI sites (E) is indicated. (Reprinted from Guarino, L. A., Gonzalez, M. A., and Summers, M. D., *J. Virol.*, 60, 224, 1986. With permission.)

plasmid p*Eco*RI-B activates expression of 39CAT (Figure 2, lane 1). Therefore, we were surprised to observe in a subsequent experiment that *Eco*RI-restricted viral DNA did not activate 39CAT expression (Figure 2, lane 3). This lack of activation could not be attributed to simple linearization of viral DNA because digestion with several other restriction enzymes did not inhibit activation of 39CAT.[6] This apparent contradiction suggests that although IE-1 efficiently induces 39K expression in a transient assay, an additional factor(s) is required for 39K expression in a viral infection.

First, we determined that the homologous regions which are rich in *Eco*RI sites were involved in 39CAT expression. Five clones of AcNPV DNA (Figure 1A) spanning the regions of homologous DNA were cotransfected with 39CAT in the presence or absence of

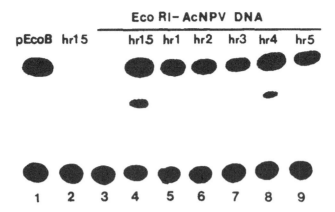

FIGURE 2.   Transient expression assays of 39CAT and homologous DNA regions. *S. frugiperda* cells were transfected and CAT assays were performed 24 h posttransfection as previously described, except that only 2 μl of extract were used for each assay.[6] The positions of input unacetylated chloramphenicol (CM) and acetylated chloramphenicol (Ac-CM) are indicated on the right side. Cells were transfected with 2 μg of p*Eco*RI-B (lane 1) or 2 μg of *Eco*RI-digested AcNPV DNA (lanes 3 to 9) plus an equimolar mixture of all five *hr* containing plasmids (lanes 2, 4) or an equimolar amount of individual plasmid containing *hr*1 (lane 5), *hr*2 (lane 6), *hr*3 (lane 7), *hr*4, (lane 8), or *hr*5 (lane 9). (Reprinted from Guarino, L. A. and Summers, M. D., *J. Virol.*, 60, 215, 1986. With permission.)

*Eco*RI-digested viral DNA (Figure 2). An equimolar mixture of all five *hr* plasmids did not stimulate CAT activity in the absence of viral DNA (Figure 2A, lane 2). CAT activity also was not detected in cotransfections of 39CAT and individual *hr* plasmids (data not shown). However, when an equimolar mixture of the *hr* plasmids was cotransfected with *Eco*RI-digested DNA, CAT activity was induced (Figure 2, lane 4). Cotransfection of the five plasmids individually with *Eco*RI-digested DNA stimulated CAT, indicating that all five *hr*s are not required for activation (Figure 2, lanes 5 to 9).

The difference between cotransfection of p*Eco*RI-B, which stimulated 39CAT, and *Eco*RI-digested DNA, which did not stimulate 39CAT, was attributed to a combination of two factors, the concentration and the conformation of the plasmid DNA. Routinely, transfections were performed using 2 μg of 39CAT plasmid and 2 μg of viral DNA or *trans*-activating plasmid. Because p*Eco*RI-B is approximately 10 map units of the viral genome, this corresponds to ten times the molar amount of *Eco*RI-B sequences cotransfected when plasmid was tested than when viral DNA was tested. Cotransfection of an amount of p*Eco*RI-B equivalent to the molar amount (23 fmol) of that fragment in 2 μg of viral DNA stimulates expression of 39CAT fourfold over background (Figure 3A). Increasing the amount of p*Eco*RI-B two- or fivefold proportionally increased the level of CAT activity induced. Ten times as much p*Eco*RI-B was not significantly greater than five times (Figure 2B, lanes 3-5). However, if p*Eco*RI-B was linearized by digestion with *Eco*RI prior to transfection, no activity was detected (data not shown), suggesting that transformation of the supercoiled plasmid DNA is more efficient. It has been reported in another system that supercoiled DNA is more efficiently expressed in transient assays than linear or nicked DNA.

Alternatively, when the amount of p*Eco*RI-B was held constant at 23 fmol, CAT activity was stimulated up to tenfold by adding a mixture of *hr* plasmids (Figure 3B). Even the addition of *hr* sequences at 0.1 times the molar amount of *Eco*RI-B (2.3 fmol) stimulated CAT activity 3.5-fold. The results of this experiment indicate that the IE-1 gene is sufficient for the induction of 39CAT in the transient assay only when added in supercoiled form in extramolar amounts. When the amount of IE-1 is equivalent to the molar amount present in 2 μg of viral DNA, then *hr* sequences are required for activation of 39K.

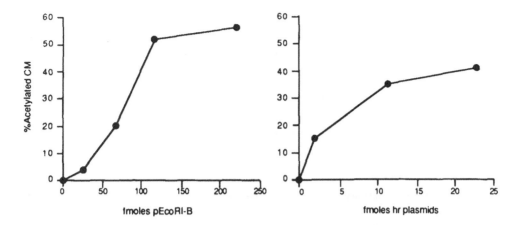

FIGURE 3.   39CAT activity as a function of p*Eco*RI-B and p*Hind*III-Q concentration. *S. frugiperda* cells were transfected and assayed as described in the legend to Figure 2. (A) Cells were cotransfected with 2 μg of 39CAT and 23, 56, 115, or 230 fmol of p*Eco*RI-B; (B) cells were cotransfected with 2 μg of 39CAT, 23 fmol of p*Eco*RI-B, and 2.3, 11.5, or 23 fmol of p*Hind*III-Q. (Reprinted from Guarino, L. A. and Summers, M. D., *J. Virol.*, 60, 215, 1986. With permission.)

Table 1
HOMOLOGOUS DNA ENHANCES EXPRESSION OF
39CAT

Transfected plasmids	CAT activity (pmol/min/106 cells)	Fold stimulation
39CAT	1.2	1
39CAT-hrl +	1173	978
39CAT-hrl −	2075	1729
39CAT-hr2 +	1306	1088
39CAT-hr2 −	2352	1960
39CAT-hr3 +	347	289
39CAT-hr3 −	243	202
39CAT-hr4*left* −	1622	1351
39CAT-hr5 +	1272	1060
39CAT-hr5 −	1259	1049

Note: *S. frugiperda* cells were transfected with 1 μg of the indicated plasmids and 0.1 μg of pIE-1. The cells were harvested and the CAT activity was measured 24 h posttransfection. Cell lysates were diluted so that less than 30% of the input chloramphenicol was acetylated. This experiment was repeated three times. (Reprinted from Guarino, L. A., Gonzalez, M. A., and Summers, M. D., *J. Virol.*, 60, 224, 1986. With permission.)

B. *hrs cis*-Activate Expression of 39CAT

The *hrs* are short repeated sequences,[9] as are several enhancer-containing sequences which activate the expression of early regulatory genes in a number of viruses.[15] However, the *hrs*-activated 39CAT expression when cotransfected on separate plasmids, while enhancers act only in *cis*. To determine whether the stimulation of 39CAT was indirectly due to *in vivo* recombination and resultant *cis*-activation of IE-1, we constructed plasmids containing the AcNPV *hrs* upstream and downstream of IE-1 in both orientations. As shown in Table 1, all five regions enhance CAT activity, although to varying degrees.

Homologous regions 1 and 2 consistently show a polar effect of orientation. It is possible that this is due to the presence of other promoters adjacent to the *hr* sequences which compete with the 39K promoter. A polar effect was also observed when the entire *Hind*III-Q fragment containing *hr*5 was cloned upstream of 39CAT.[16] However, this polarity was not evident with the *Mlu*I fragment of *Hind*III-Q which contains only *hr* sequences.

The enhancement seen with *hr*3 is significantly lower than that observed with the other enhancers. This result is unexpected as the sequence for *hr*3 is not significantly different from the other regions, except for the large number of small *Eco*RI fragments. It seems unlikely that this would account for the poor level of enhancement as there are three larger fragments, and as discussed below, a single fragment is sufficient for enhancer function.

The results with *hr4left* indicate that a single *Eco*RI minifragment stimulates as efficiently as regions containing multiple repeats. This indicates that the AcNPV *hr* enhancers are similar to other enhancers[15] which contain repeated sequences, although the repeats are not essential for enhancer activity.

C. Nucleotide Sequence of AcNPV Homologous DNA

The sequencing strategy for the five homologous regions is shown in Figure 1, and the number of nucleotides between adjacent *Eco*RI sites is indicated. The complete sequence of the five *hr*s has previously been presented by Guarino et al.[17] The homology of the repeated sequences in the *hr*s is demonstrated in Figure 4. In this analysis, gaps have been inserted for optimal alignment of homologous sequences, and some bases from the longer *Eco*RI minifragments have been omitted. Homologous sequences extending to the left and right of the first and last *Eco*RI sites are also included. The number of homologous bases beyond the *Eco*RI sites is variable in both directions, and there is apparently no sequence common to all 5' or 3' ends of the *hr*s.

The 60 base pairs (bp) surrounding the *Eco*RI sites are highly conserved. This conserved region is separated by nonhomologous sequences of variable length between 12 and 155 bp. The percent of homology between the consensus sequences for each region and the individual *Eco*RI minifragments is 80% for *hr*1, *hr*2, and *hr*3; 65% for *hr4right*; and 87% for *hr*5. The low homology for *hr4right* is a reflection of the large difference in size between the 215-bp minifragment and the two smaller fragments. Both the conserved and nonconserved sequences of the *hr*s are rich in adenine (A) and thymine (T). The average A + T content of the *hr*s is 67%, significantly higher than the average 58% A + T content of the viral genome.[18]

The homology between the consensus sequences of each region is also shown in Figure 4. The homology of the consensus sequences is 88% when two regions (*hr*1 and *hr*3) are compared in the genome antisense orientation. The homology between the consensus sequences was only 75% (not shown) when all regions were compared in the genome sense orientation. This indicates that there is a polarity to the sequence. However, the function of the AcNPV enhancers apparently is not polar (Table 1).

The most highly conserved feature of the *hr*s is a 26-bp imperfect, inverted repeat (palindrome). The consensus sequences for the five palindromes is shown in Figure 5. The *Eco*RI sites form the core of this palindrome. In four regions, there is a single-base pair mismatch in the consensus palindrome. In *hr*5, there is a 2-bp mismatch in the consensus sequence, and the conserved palindrome is 28 bp. It is interesting to note that while the sequence of the individual palindromes may differ from the consensus sequence, there is always a mismatch in the same location. There are no instances of a perfect 26-bp palindrome, suggesting a functional significance for the mismatch.

IV. DISCUSSION

The results presented here demonstrate that all five *hr*s enhanced expression of CAT. Although the *hr* sequences were not required for efficient expression of 39CAT in a transient

hr1

```
201  TCGATTGTGT TTTACAAGTA GAATTCTACC CGTAAAGCGA GTTTAGTTTT (22) .......... .......... .......... ..TATAATGA CATCATCCC
269  CTGATTGTGT TTTACAAGTA GAATTCTATC CGTAAAGCGA GTTCAGTTTT (59) AAGTTATGAG CCGTGTGCAA AACATGAGAT AAGTTTATGA CATCATCCA
428  CTGATCGTGC GTTACAAGTA GAATTCTACT CGTAAAGCCA GTTC...... .GGTTATGAG CCGTGTGCAA AACATGACAT CAGCTTATGA C.TCATAC.
518  TTGATTGTGT TTTACGCGTA GAATTCTACT CGTAAAGCGA GTTC...... .GGTTATGAG CCGTGTGCAA AACATGACAT CAGCTTATGA .GTCATAA.
619  TTAATCGTGC GTTACAAGTA GAATTCTACT CGTAAAGCGA GTT        .GGTTATGAG CCGTGTGCAA AACATGACAT CAGCTTATGA
con  ttgattgtgt tttacaagta gaattctact cgtaaagcga gttc------ .ggttatgag ccgtgtgcaa aacatgacat cagcttatga catcat-c.
```

hr2

```
101  TTTACGAGTA GAATTCTACG TGTAAAAACAT AATCAAGAGA TGATGTCATT TGTTTTTCAA ( 8) TCAAGAAATG ATGTCATTTG TTTTTCAAAA CTGAACTGGC
209  TTTACGAGTA GAATTCTACT TGTAAAAACAC AATCGAGAGA TGATGTCCAT ATTTTGCACA (63) GTCATTGGAT GAGTCATTTG TTTTTCAAAA CTAAACTCGC
311  TTTACGAGTA GAATTCTACT TGTAAAAACAC AATCAAGGGA TGATGTCATT .......... .ATACAAATG ATGTCATTTG TTTTTCAAAA CTAAACTCGC
459  TTTACGGGTA GAATTCTACT TGTAAAAACAG CAACTCGAGG GATGATGTCA TCCTTTACTC (14) TGTTATGTAT GACTCATTTG TTTTTCAAAA CTAAACTCGC
575  TTTACGAGTA GAATTCTACT TGTAACGCAC GATCAAGGGA TGATGTCATT TATTTGTGCA (39) TCCAAATAAT GACTCATTTG .TTTTCAAAA CTGAACTCGC
712  TTTACGAGTA GAATTCTACT TGTAAAAACAC AATCAAGGGA TGATGTCATT TT........ ...CAAAATG ATGTCATTTG TTTTTCAAAA CTAAACTCGC
802  TTTACGAGTA GAATTCTACT TGTAAAAACAC AATCAAGGGA TGATGTCATT TT........ .....AAAAA TGATCATTTG TTTTTCAAAA CTAAACTCGC
889  TTTACGAGTA GAATTCTACG TGTAAAAACAC AATCAAGGGA TGATGTCATT T
con  tttacgagta gaattctact tgtaaaaacac aatcaaggga tgatgtcatt ttttt..... ---aaaaaat g-gtcatttg tttttcaaaa ctaaactcgc
```

hr3

```
1    TTTACGCGTA GAATTCTACT TGTAAAGCAA GTTAAAATAA GCCGTGTGCA AAAATGACAT CAGACAAATG (48) CATCAGCTTA TGACTAAATAA TTGATCGTGC
146  GTTACAAGTA GAATTCTACT CGTAAAGCGA GTTTAGTTTT GAAAA.CAAA TGAGTCATCA ATTAAACATG (13) ATAAAGGATG ACATCATCCA CTAATCGTGC
257  GTTACAAGTA GAATTCTACT CGTAAAGCGA GTTCGGTTTT GAAAAACAAA TGACATCATT .......... .......... ........TC TTGATTATGT
329  TTTACAAGTA GAATTCTACT CGTAAAGTAT GTTCAGTTT. AAAAAACAAA TGACATCATT .......... .......... ..ATCATTTC TTGATTATGT
417  TTTACAAGTA GAATTCTACT CGTAAAGCAA GTTTAGTTTT AAAAAACAAA TGACATCA.. .......... .......... ....TCTC TTGATTATGT
489  TTTACAAGTA GAATTCTACT CGTAAAGCGA GTTTAGTTTT GAAAAACAAA TGACATCA.. .......... .......... ....TCTC TTGATTATGT
561  TTTACAAGTA GAATTCTACT CGTAAAGCGA GTTTAGTTTT GAAAAACAAA TGACA..... .......... .......... ...TCATCCC TTGATCATGC
633  GTTACAAGTA GAATTCTACT CGTAAAGCGA G.TGAATTTT GA
con  tttacaagta gaattctact cgtaaagcga gtttagtttt gaaaaacaaa tgacatca.. .......... .......... ..atcatctc ttgattatgt
```

hr4

```
1    AACTGGCTTT ACGAGTAGAA TTCTACTTGT AAAACACAAT CAAGAAATGA TGTCATTTTT GTACGTGATT ATAAACATGT TTAAACATGG (115) TCATCGTACA
216  AACTCGCTTT ACGAGTAGAA TTCTACTTGT AAAACACAAT CGAGGGATGA TGTCATTTGT AGAATGATGT CATTTGTTTT TTCAAAACCG ..........
306  AACTCGCTTT ACGAGTAGAA TTCTACTTGT AAAACACAAT CGAGGGATGA TGTCATTTGT AGAATGATGT CAT....... .......... ....CGTACA
385  AACTCGCTTT ACGAGTAGAA TTCTAGT.GT AAAACAC
con  aactcgcttt acgagtagaa ttctacttgt aaaacacaat cgagggatga tgtcatttgt agaatgatgt cat....t.t tt.aa.a..g ....cgtaca
```

hr5

```
69   TTAAAATTGA ACTCGGCTTT ACGAGTAGAA TTCTACGCGT AAAACACAAT CAAGT.ATGA ( 7) .......... .....GCTGA TGTCATGTTT TGCACACGGC
165  TCATAACCGA ACT.GGCTTT ACGAGTAGAA TTCTACGTGT AACGCACGAT CGAGTGGATG (10) GTTTTTCAAA TCGA.GATGA TGTCATGTTT TGCACACGGG
273  CTCATAAACT .....GCTTT ACGAGTAGAA TTCTACGTGT AACGCACGAT CGATTGATGA (11) TTTTGCAATA TGATATCATA CAATATGACT CATTTGTTTT
376  TCAAAACCGA ACTTGA.TTT ACGGGTAGAA TTCTACTCGT AAAGCACANT CAA....... .......... ...AAAGATGA TGTCATTTGT TTT.......
449  TCAAAACTGA ACTCGGCTTT ACGAGTAGAA TTC.ACGTGT AAAACACAAT CAAGAAATGA ( 8) GTTATAAAAA TAAAAGCTGA TGTCATGTTT TGCACATGGC
558  TCATAACTAA ACTC.GCTTT ACGGGTAGAA TTCTACGCGT AAAACA
con  tcaaaactga actcggcttt acgagtagaa ttctacgcgt aaagcacaat caagt.gatga -tt--.-a-a t.aaagctga tgtcatgttt tgcaca-ggc
```

CONSENSUS

```
1-   GAATTCTACT TGTAAAAACAC AATCAA.GGA TGATGTCATA AGCTGATGTC ATGTTTTGCA CACGGCTCAT TAACC.AAAA CTGAACTCGC TTTACGAGTA
2-   GAATTCTACT TGTAAAACAC AATCAAGGGA TGATGTCATT TTTTT..... .TTAAAAAA- --GTCATTTG TTTTTCAAAA CTAAACTCGC TTTACGAGTA
3-   GAATTCTACT TGTAAAACAT AATCAAGAGA TGAT...... .......... ........TT GTTGTCTTTG TTTTTCAAAA CTAAACTCGC TTTACGAGTA
4-   GAATTCTACT TGTAAAAACAC AATCGAGGGA TGATGTCATT TGTAGAATGA TGTCAT.... .........TT TTAAAGACGT ACAAACTCGC TTTACGAGTA
5-   GAATTCTACG TGTAAAGCAC AATCAAGTGA TGA-TT--.- A-AT.AAAGC TGATGTCATG TTTTGCACTG GCTCAAAACT GAACTCGGCC TTTACGAGTA
con  gaattctact tgtaaaaacac aatcaaggga tgatgtcatt -gttga-.g- --ttat.-., .tttgctttg tttttcaaaa ctaaactcgc tttacgagta
```

FIGURE 4. Alignment of homologous sequences. The nucleotide sequences of the five *hr*s are presented with gaps inserted for optimal alignment of homologous sequences. Some bases from the longer fragments are not shown, the number of bases is indicated in parentheses. *Eco*RI restriction sites are underlined. The numbers on the left refer to the base numbers in Figure 2. The consensus sequence (con) is indicated for each region; where the consensus can be more than one base, a dash is indicated. The consensus sequences for each homologous region are aligned. In some cases the dashes have been changed to the base which best fits the overall consensus. The sequences for *hr*1 and 3 are presented in the opposite (−) orientation as the standard AcNPV genetic map; the other *hr*s are shown in the standard (+) orientation. (Reprinted from Guarino, L. A., Gonzalez, M. A., and Summers, M. D., *Jr. Virol.*, 60, 224, 1986. With permission.)

assay, the data suggests that *hr*s may be essential for 39K expression in a viral infection. There was no CAT activity induced by AcNPV DNA which had been previously digested with *Eco*RI prior to transfection. This result is apparently specific for *Eco*RI; digestion with several other restriction enzymes prior to transfection did not affect 39CAT expression.[6] Furthermore, CAT activity was not observed upon cotransfection of 39CAT and pIE-1, when the amount of IE-1 sequences added were equivalent to the molar amount of IE-1 in transfections with viral DNA. Only when extramolar amounts of pIE-1 were transfected was the requirement for *hr* sequences bypassed.

Two- to tenfold stimulation of CAT activity was observed when the equimolar amounts of the *hr* plasmids and 39CAT were cotransfected on separate plasmids, and CAT activity

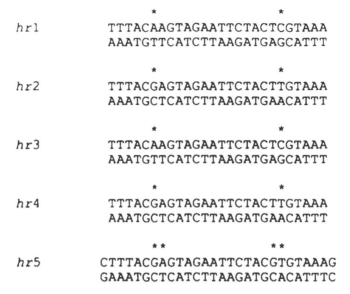

FIGURE 5.   Highly conserved palindromes in *hr* DNA. Mismatches in the palindromes are indicated by an asterisk. (Reprinted from Guarino, L. A., Gonzalez, M. A., and Summers, M. D., *J. Virol.*, 60, 224, 1986. With permission.)

was further increased 200- to 1000-fold when the *hr* sequences were cloned upstream of 39CAT. The stimulation observed with separate plasmids may be due to a true *trans* effect of the *hr*s. However, a more probable explanation is that the *trans* activity is due to *in vivo* recombination and resultant *cis* activation. If only one percent of the p39CAT and *hr* plasmids molecules recombined and those recombinants were expressed at a 1000 times the level of the input molecules, an apparent tenfold stimulation would be observed.

The AcNPV enhancer is similar to vertebrate viral enhancers with respect to orientation independence and ability to function at a distance. The AcNPV enhancers differ significantly from the enhancers of the vertebrate DNA viruses with respect to their physical location in the viral genome. Many vertebrate viral genomes contain a single enhancer located within a few hundred base pairs of major immediate early genes.[15] Apparently the function of these enhancers is to *cis*-activate transcription of regulatory genes which in turn *trans*-activate delayed early transcription. However, the AcNPV genome contains multiple enhancers interspersed throughout the viral genome. The immediate early regulatory gene, IE-1, is at least 3 kb from the nearest *hr*. Transcription of IE-1 was not affected by linked *hr* sequences, whereas the expression of a delayed early gene was enhanced 1000-fold.

REFERENCES

1. **Carstens, E. B., Tjia, S. T., and Doerfler, W.,** Infection of *Spodoptera frugiperda* cells with *Autographa californica* nuclear polyhedrosis virus, *Virology,* 99, 386, 1979.
2. **Maruniak, J. E. and Summers, M. D.,** *Autographa californica* nuclear polyhedrosis virus phosphoproteins and synthesis of intracellular proteins after virus infection, *Virology,* 109, 25, 1981.
3. **Adang, M. J. and Miller, L. K.,** Molecular cloning of DNA complementary to mRNA of the baculovirus *Autographa californica* nuclear polyhedrosis virus: location and gene products of RNA transcripts found late in infection, *J. Virol.,* 44, 782, 1982.

4. **Esche, H., Lubbert, H., Siegmann, B., and Doerfler, W.,** The translation map of the *Autographa californica* nuclear polyhedrosis virus (AcNPV) genome, *EMBO J.*, 1, 1629, 1982.
5. **Vlak, J. M., Smith, G. E., and Summers, M. D.,** Hybridization selection and in vitro translation of *Autographa californica* nuclear polyhedrosis virus, *J. Virol.*, 40, 762, 1981.
6. **Guarino, L. A. and Summers, M. D.,** Functional mapping of a *trans*-activating gene, *J. Virol.*, 57, 563, 1986.
7. **Guarino, L. A. and Summers, M. D.,** Nucleotide sequence and enhancer function of a baculovirus regulatory gene, *J. Virol.*, 61, 2091, 1987.
8. **Fuchs, L. Y., Woods, M. S., and Weaver, R. F.,** Viral transcription during *Autographa californica* nuclear polyhedrosis virus infection: a novel RNA polymerase induced in infected *Spodoptera frugiperda* cells, *J. Virol.*, 48, 641, 1983.
9. **Cochran, M. A. and Faulkner, P.,** Location of homologous DNA sequences interspersed at five regions in the baculovirus AcMNPV genome, *J. Virol.*, 45, 961, 1983.
10. **Kuzio, J. and Faulkner, P.,** Regions of repeated DNA in the genome of *Choristoneura fumiferana* nuclear polyhedrosis virus, *Virology*, 139, 185, 1984.
11. **Maniatis, T., Fritsch, E. F., and Sambrook, J.,** *Molecular Cloning*, Cold Spring Harbor Laboratory, Cold Spring Harbor, NY.
12. **Messing, J.,** New M13 vectors for cloning, in *Methods of Enzymology*, Vol. 101, Colowick, S. P. and Kaplan, N. D., Eds., Academic Press, New York, 1983, 20.
13. **Sanger, F., Nicklen, S., and Coulson, A. R.,** DNA sequencing with chain terminating inhibitors, *Proc. Natl. Acad. Sci. U.S.A.*, 74, 5463, 1977.
14. **Devereaux, J., Haeberli, P., and Smithies, O.,** A comprehensive set of sequence analysis programs for the VAX, *Nucleic Acids Res.*, 12, 387, 1984.
15. **Gluzman, Y. and Shenk, T., Eds.,** *Enhancers and Eukaryotic Gene Expression*, Cold Spring Harbor Laboratory, Cold Spring Harbor, NY, 1983.
16. **Guarino, L. A. and Summers, M. D.,** Interspersed homologous DNA of *Autographa californica* nuclear polyhedrosis virus enhances delayed-early gene expression, *J. Virol.*, 60, 215, 1986.
17. **Guarino, L. A., Gonzalez, M. A., and Summers, M. D.,** Complete sequence and enhancer function of the homologous DNA regions of *Autographa californica* nuclear polyhedrosis virus, *J. Virol.*, 60, 224, 1986.
18. **Vlak, J. M. and Odink, K. G.,** Characterization of *Autographa californica* nuclear polyhedrosis virus deoxyribonucleic acid, *J. Gen. Virol.*, 44, 333, 1979.

Chapter 23

PROCESSING OF THE HEMAGGLUTININ OF INFLUENZA VIRUS EXPRESSED IN INSECT CELLS BY A BACULOVIRUS VECTOR

Kazumichi Kuroda, Charlotte Hauser, Rudolf Rott, Walter Doerfler, and Hans-Dieter Klenk

TABLE OF CONTENTS

I. INTRODUCTION

Like other integral membrane proteins, the hemagglutinin (HA) as the major glycoprotein of the influenza virus envelope is translated at the rough endoplasmic reticulum and transported from there through the Golgi apparatus to the plasma membrane. In the course of transport, the HA undergoes posttranslational modifications, including removal of the signal peptide, attachment of N-glycosidic oligosaccharide side chains,[1] acylation,[2] and posttranslational proteolytic cleavage of the precursor HA into the fragments HA_1 and HA_2, which is necessary for the realization of the fusion activity.[3-5]

Recently, an expression system was developed in insect cells using *Autographa californica* nuclear polyhedrosis virus (AcNPV) as a vector. This system proved to be very valuable, because of the efficient polyhedrin promoter of AcNPV, and has been successfully applied for the expression of several secretory proteins, such as human β-interferon,[6] β-galactosidase from *Escherichia coli*,[7] human c-myc protein,[8] and human interleukin 2.[9]

The first membrane protein expressed by an AcNPV vector was the HA of influenza virus.[10,11] Synthesis of the HA provides valuable information on the capacity of the system to exert the complex processing steps required for the biological function of such proteins.

II. MATERIALS AND METHODS

The insect cell line from *Spodoptera frugiperda* was propagated in modified TC-100 medium[12] containing 10% fetal calf serum. The methods employed in virus infection and in the purification of extracellular AcNPV have been detailed previously.[12,13] Cotransfection of *S. frugiperda* cells with DNA from AcNPV and a vector containing the HA gene of fowl plague virus, an avian influenza virus (pAc-HA651), was done essentially as described by Smith and co-workers.[14] Recombinant virus was selected by a two-step procedure involving adsorption of transfected cells to erythrocytes and plaque purification.[10] Polyadenylated RNA was extracted from the cytoplasm and purified by oligo-(dT)-chromatography as described.[15]

III. RESULTS

A. Construction of Recombinant Virus

The HA gene of the fowl plague virus mutant ts-651 was cloned into the *Pst*I site of the plasmid pUC 8, using the dG-dC tailing procedure. After the removal of the tails, the HA gene was cloned into the *Bam*HI site of pAc 373.[10] One clone was chosen and the flanking region of the HA gene was sequenced (Figure 1). The nucleotide sequence data proved that the HA gene had been correctly cloned into the *Bam*HI site of pAc 373. To transfer the HA gene into the AcNPV genome, *S. frugiperda* cells were cotransfected with pAc-HA651 and the genomic DNA of the authentic virus. The recombinant AcNPV was selected as described in Section II.

B. Transcription of the Hemagglutinin Gene in *S. frugiperda* Cells

Polyadenylated RNA was isolated from *S. frugiperda* cells infected with authentic virus, recombinant virus, or from mock infected cells, and analyzed by the Northern blotting technique. The blot was hybridized with a HA-specific cDNA probe. Only in recombinant virus infected cells was hybridization observed. There were two main bands corresponding to the molecular sizes of 2.8 and 1.8 kb, respectively (Figure 2). These values agreed well with what had to be expected from the constructs of pAc-HA651 (Figure 1). The HA gene insert (1.74 kb) started 80b downstream from the transcriptional initiation site and ended ≈1 kb upstream from the transcriptional termination site of the deleted polyhedrin gene. As both genes have the termination and polyadenylation signal, two kinds of transcripts are

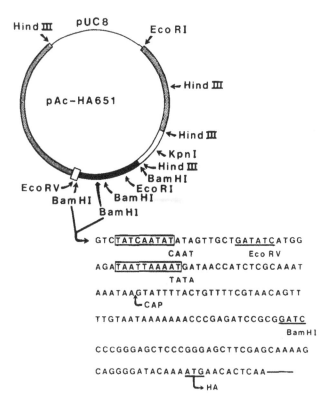

FIGURE 1. Construction of pAc-HA651. Fowl plague virus hemagglu-tinin gene, polyhedrin, and other AcNPV sequences are indicated by solid, blank, and dotted bars, respectively. Polyhedrin promoter is indicated by rectangle. The nucleotide sequence of 5′ flanking region of hemagglutinin gene is also shown.

expected, i.e., one starting from the start site of the polyhedrin gene and ending at the termination signal of the hemagglutinin gene ($\approx$1.8 kb), and another one starting from the same site and ending at the termination signal of the polyhedrin gene ($\approx$2.8 kb).

C. Expression and Processing of the Influenza Hemagglutinin in *S. frugiperda* Cells

In homogenates of recombinant infected cells, hemagglutinating activity could be detected starting at $\approx$24 h after inoculation, and the titer increased continuously up to $\approx$96 h, when most of the cells deteriorated. The infected cells showed also hemadsorption activity. At $\approx$48 h after infection, each cell was heavily loaded with erythrocytes (Figure 3). In contrast to the hemagglutinating activity, the amount of adsorbed erythrocytes decreased after 48 h, probably due to the disintegration of the cells. Neither authentic virus nor mock infected cells displayed hemagglutination or hemadsorption activities (Figure 3). Hemolytic activity dependent on acidic pH, another biological activity of the HA, was also detected in ho-mogenates of cells infected with recombinant virus, and the pH dependence of this activity was similar to that of authentic HA. The HA produced in *S. frugiperda* cells was also analyzed by the immunoblotting technique. It was clearly demonstrated that the HA was cleaved in *S. frugiperda* cells, even though to a lesser extent than in influenza virus-infected MDCK cells (Figure 4).

Figure 4 demonstrates also that the HA has a lower molecular weight when derived from *S. frugiperda* cells. This difference is due to variations in the carbohydrate content. Incubation with endoglycosidase H and glycopeptidase F indicated that the HA is, indeed, glycosylated

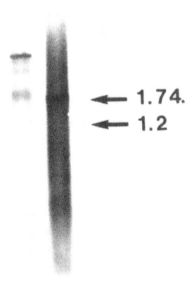

FIGURE 2.   Analysis of mRNA by Nothern blotting. *S. frugiperda* cells were mock-infected (Mock) or were infected with authentic (Wild) or recombinant AcNPV (Rec) for 3 d, and cytoplasmic mRNA was extracted. mRNA and genomic RNA (FPV) of fowl plague virus were separated on an agarose gel containing formaldehyde and blotted to nylon membrane. The blot was hybridized with ^{32}P-labeled hemagglutinin cDNA.

in *S. frugiperda* cells (Figure 5). After treatment with glycopeptidase F, which removes carbohydrate side chains of both the oligomannosidic and the complex type, a single HA band was observed representing the polypeptide completely stripped of its oligosaccharides. This band was also seen when HA obtained from infected MDCK cells was subjected to glycopeptidase F treatment (data not shown). Incubation with endoglycosidase H, which cleaves only oligomannosidic side chains, also yielded the carbohydrate-free HA and an additional band migrating between this and the untreated HA. Similar observations were made on HA_1 and HA_2. These findings demonstrate that the HA obtained from *S. frugiperda* cells contains a mixture of oligomannosidic- and of endoglycosidase-H-resistant side chains, the latter ones being smaller in size than the complex oligosaccharides found on HA synthesized in vertebrate cells.

D. Immunization of Chickens with Recombinant Hemagglutinin

To test the immune response to recombinant HA, chickens were immunized. The immunized animals produced hemagglutination-inhibiting antibodies at titers of 1:64 to 1:128. The sera neutralized infectivity of fowl plague virus with indices ≥ 500. The immunized chickens survived a challenge infection with 10^4 p.f.u. of fowl plague virus 3 weeks after immunization without showing signs of fowl plague, while unprotected control animals died 2 d after infection, as expected.

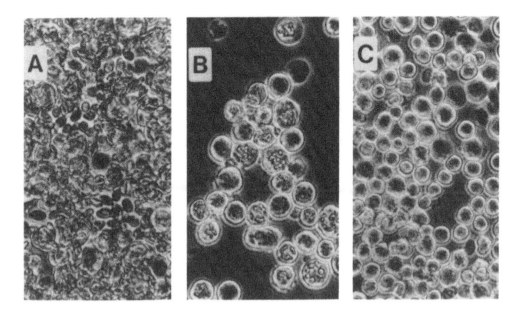

FIGURE 3. Hemadsorption of *S. frugiperda* cells after infection with recombinant AcNPV. At 2 d after infection with recombinant AcNPV (A), after infection with authentic AcNPV (B), or after mock infection (C), *S. frugiperda* cells were analyzed for hemadsorption.

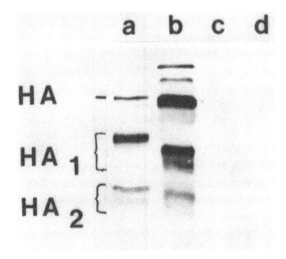

FIGURE 4. Analysis of hemagglutinin by immunoblotting. MDCK cells were infected with fowl plague virus (a) for 5 h. *S. frugiperda* cells were infected with recombinant (b), authentic AcNPV (c), or were mock-infected (d) for 3 d. The polypeptides were separated by SDS-polyacrylamide gel electrophoresis, and then blotted to a nitrocellulose membrane. The blot was analyzed with antihemagglutinin antibody. Samples derived from 10^5 cells were loaded onto each lane of the gel.

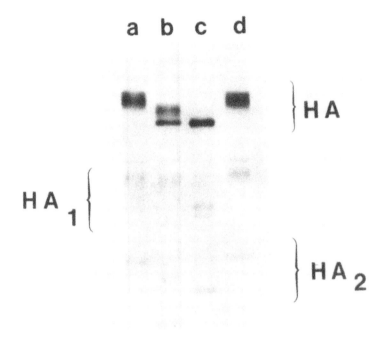

FIGURE 5.   Analysis of the carbohydrates of the hemagglutinin by digestion with endo-glycosidase H and glycopeptidase F. *S. frugiperda* cells were infected with recombinant AcNPV for 2 d. The infected cells were labeled by a 3-h pulse with ^{35}S-methionine, and the hemagglutinin was immunoprecipitated. The precipitated hemagglutinin was analyzed by SDS-polyacrylamide gel electrophoresis directly (a), or after 16-h incubation with endog-lycosidase H (b), glycopeptidase F (c), or PBS (d).

IV. DISCUSSION

We have constructed recombinant AcNPV which contained the influenza virus HA gene downstream of the polyhedrin promoter. In cells infected with the recombinant, the HA gene was transcribed and translated. The synthesized HA is able to induce hemadsorption, cause hemolysis, and react with HA-specific antibodies. Moreover, the recombinant HA can protect animals against the challenge by fowl plague virus. Thus, in its biological activities the HA produced by recombinant virus strikingly resembles authentic fowl plague virus HA.

Our data show that the recombinant HA is membrane-bound in *S. frugiperda* cells and that it is transported to the cell surface. It is also shown that the HA undergoes posttrans-lational proteolytic cleavage and glycosylation in insect cells. The observation that the recombinant HA has hemolytic activity indicates that the endoprotease involved in cleavage attacks the correct peptide bond since hemolytic capacity is expressed only under these conditions.[16] The digestion of the HA with endoglycosidase H and glycopeptidase F clearly shows the attachment of sugar chains and the processing of these chains. The recombinant HA is partially resistant to the endoglycosidase H. This result agrees well with our previous report showing the presence of five complex and two oligomannosidic sugar chains in fowl plague virus HA.[1] It is known that N-glycosidic sugars are trimmed in insect cells to an endoglycosidase H-resistant oligosaccharide containing three mannose residues, but not elongated to complex sugars.[17] This might explain the difference in molecular size between authentic and recombinant HA. It is known that fowl plague virus HA requires its glycosyl groups for full biological activity. The glycosylating capacity of *S. frugiperda* cells appears to be sufficient to meet this requirement.

REFERENCES

1. **Keil, W., Geyer, R., Dabrowski, U., Niemann, H., Stirm, S., and Klenk, H.-D.,** Carbohydrates of influenza virus. Structural elucidation of the individual glycans of the FPV hemagglutinin by two-dimensional ¹H n.m.r. and methylation analysis, *EMBO J.,* 4, 2711, 1985.
2. **Schmidt, M. F. G.,** Acylation of viral spike glycoproteins: a feature of enveloped RNA viruses, *Virology,* 116, 327, 1982.
3. **Huang, R. T. C., Rott, R., and Klenk, H.-D.,** Influenza viruses cause hemolysis and fusion of cells, *Virology,* 110, 243, 1981.
4. **Klenk, H.-D., Rott, R., Orlich, M., and Blödorn, J.,** Activation of influenza A viruses by trypsin treatment, *Virology,* 68, 426, 1975.
5. **Lazarowitz, S. G. and Choppin, P. W.,** Enhancement of the infectivity of influenza A and B viruses by proteolytic cleavage of the hemagglutinin polypeptide, *Virology,* 68, 440, 1975.
6. **Smith, G. E., Summers, M. D., and Fraser, M. J.,** Production of human beta interferon in insect cells infected with a baculovirus expression vector, *Mol. Cell. Biol.,* 3, 2156, 1983.
7. **Pennock, G. D., Shoemaker, C., and Miller, L. K.,** Strong and regulated expression of *Escherichia coli* β-galactosidase in insect cells with a baculovirus vector, *Mol. Cell. Biol.,* 4, 399, 1984.
8. **Miyamoto, C., Smith, G. E., Farrell-Towt, J., Chizzonite, R., Summers, M. D., and Ju, G.,** Production of human c-myc protein in insect cells infected with a baculovirus expression vector, *Mol. Cell. Biol.,* 5, 2860, 1985.
9. **Smith, G. E., Ju, G., Ericson, B. L., Moschera, J., Lahm, H.-W., Chizzonite, R., and Summers, M. D.,** Modification and secretion of human interleukin 2 produced in insect cells by a baculovirus expression vector, *Proc. Natl. Acad. Sci. U.S.A.,* 82, 8404, 1985.
10. **Kuroda, K., Hauser, C., Rott, R., Klenk, H.-D., and Doerfler, W.,** Expression of the influenza virus haemagglutinin in insect cells by a baculovirus vector, *EMBO J.,* 5, 1359, 1986.
11. **Possee, R. D.,** Cell-surface expression of influenza virus haemagglutinin in insect cells using a baculovirus vector, *Virus Res.,* 5, 43, 1986.
12. **Carstens, E. B., Tjia, S. T., and Doerfler, W.,** Infection of *Spodoptera frugiperda* cells with *Autographa californica* nuclear polyhedrosis virus. I. Synthesis of intracellular proteins after virus infection, *Virology,* 99, 386, 1979.
13. **Tjia, S. T., Carstens, E. B., and Doerfler, W.,** Infection of *Spodoptera frugiperda* cells with *Autographa californica* nuclear polyhedrosis virus. II. The viral DNA and the kinetics of its replication, *Virology,* 99, 399, 1979.
14. **Smith, G. E., Fraser, M. J., and Summers, M. D.,** Molecular engineering of the *Autographa californica* nuclear polyhedrosis virus genome: deletion mutations within the polyhedrin, *J. Virol.,* 46, 584, 1983.
15. **Maniatis, T., Fritsch, E. F., and Sambrook, J.,** *Molecular Cloning: A Laboratory Manual,* Cold Spring Harbor Laboratory Press, Cold Spring Harbor, NY, 1982.
16. **Garten, W., Bosch, F. X., Linder, D., Rott, R., and Klenk, H.-D.,** Proteolytic activation of the influenza virus hemagglutinin: the structure of the cleavage site and the enzymes involved in cleavage, *Virology,* 115, 361, 1981.
17. **Hsieh, P. and Robbins, P. W.,** Regulation of asparagine-linked oligosaccharide processing: oligosaccharide processing in *Aedes albopictus* mosquito cells, *J. Biol. Chem.,* 259, 2375, 1984.

Index

INDEX

Printed and bound by CPI Group (UK) Ltd, Croydon, CR0 4YY

22/10/2024

01777600-0006